THE

CAMPAIGN AND BATTLE

OF

GETTYSBURG

PREPARED FOR THE USE OF THE CADETS OF THE UNITED STATES MILITARY ACADEMY, WEST POINT, N. Y.

By

COLONEL G. J. FIEBEGER

Professor of Engineering

U. S. M. A.

CHAPTER I.

CHAPTER II.

CHAPTER III.

CHAPTER IV.

CHAPTER V.

CHAPTER VI.

CHAPTER VII.

CHAPTER IX.

CHAPTER I.

MILITARY SITUATION AT THE OPENING OF THE GETTYSBURG CAMPAIGN.

CONFEDERATE FORCES.—On May 31, 1863, the total strength of the Confederate forces in the East, officers and men, was as follows:

PRESENT FOR DUTY.

Army of Northern Virginia—General Lee....	74,468	W. R. Vol. 40, p. 846
Dep't Western Virginia—General Sam Jones.	6,422	" " " 40, " 846
" of Richmond—General Elzey..........	8,464	" " " 26, " 1,086
" North Carolina—General D. H. Hill...	23,536	" " " 26, " 1,086
" South Carolina, etc.—Gen. Beauregard.	22,040	" " " 20, " 964
Valley District—General John D. Imboden...	2,100	Battles and Leaders, Vol. 3, p. 422.
Total..................................	137,030	

The *Army of Northern Virginia* was encamped on the south bank of the Rappahannock, in the vicinity of Fredricksburg, Virginia, and its cavalry was guarding all the fords of the Rappahannock as far west as the Orange and Alexandria railroad. General Longstreet, who had been at Suffolk, Virginia, during the Chancellorsville campaign, with the divisions of Hood and Pickett, had returned to the army. Hood was at this time near Verdiersville, about half way between Chancellorsville and Orange Court House; Pickett was at Hanover Junction watching the Union troops recently landed at West Point on the York river.

The *Department of Western Virginia* was the territory along the Tennessee and Virginia railroad in the southwestern part of the state. Its troops were to defend the railroad and the important salt works at Saltville, Virginia, against raids made from West Virginia and Kentucky. In this department were four small brigades of infantry, Jenkins' cavalry brigade, a few batteries of artillery and a few detached companies of infantry.

The *Department of Richmond* was the territory between the Army of Northern Virginia and the James river. Its troops were to man the defenses of Richmond and to protect that capital from the Union forces operating on the peninsula between the York and James rivers. In this Department were Wise's brigade on the Chickahominy river, Pettigrew's brigade near Hanover, and the troops in the defenses of the city and in the batteries along the James river.

The *Department of North Carolina* embraced that part of the state of Virginia south of the James river, and the entire state of North Carolina. Its troops were to resist invasion by the Union troops operating from Suffolk, Va., and also to protect North Carolina from the Union troops operating from Newberne, N. C.

In this department were seven field brigades; one at Wilmington, N. C., three between Newberne, N. C. and the Wilmington and Weldon railroad, one at Petersburg, Va., one between Petersburg and Suffolk, Va., and one, Davis' brigade, was on its way to join the Army of Northern Virginia. In addition to these, there was a small cavalry force in eastern North Carolina, and a permanent artillery garrison in the defenses of Wilmington.

All the troops in the departments of Richmond and North Carolina were under the orders of General Lee; however, he could not order them out of these departments without the consent of the President or the Secretary of War.

The *Department of South Carolina, Georgia and Florida* was made up of these states. Of the troops given in the table about 12,500 were in South Carolina; 6,600 in Georgia and the remainder in Florida. These troops were to defend Charleston, which was threatened by land and sea, and also to protect the coastal territory from raids made from the different harbors occupied by the Northern troops. Nearly all the organized field brigades of this department had been recently sent either to General Joseph E. Johnston in Mississippi, or to General D. H. Hill in North Carolina.

In the *Valley District* General Imboden had a mixed command of cavalry and mounted infantry organized for raids along the line of the Baltimore and Ohio railroad; its headquarters when not on a raid were in the upper Shenandoah Valley.

UNION FORCES.—The total strength of the Union forces in the same territory at this time was as follows:

PRESENT FOR DUTY.

Army of the Potomac—General Hooker (Infantry, Cavalry and Field Artillery)—	100,780	W. R. Vol. 40, p. 574
Dep't of Washington—General Heintzleman	36,647	" " " 40, " 586
Middle Department—General Schenck	32,247	" " " 40, " 589
Dep't of Virginia—General Dix	32,397	" " " 26, " 733
Dep't of North Carolina—General Foster	15,699	" " " 26, " 736
Dep't of the South—General Hunter	15,745	" " " 20, " 461
Total	233,515	

The *Army of the Potomac* was encamped on the north bank of the Rappahannock opposite Fredericksburg and its cavalry was guarding the upper Rappahannock as far west as the Orange and Alexandria railroad. Because of the expiration of the time of service of many of the regiments from New York, New Jersey and Pennsylvania, several of the army corps were being reorganized. The cavalry, which had lost many of its horses in the Chancellorsville campaign, was securing remounts.

The *Department of Washington* comprised the District of Columbia with the national capital, Washington. Most of the troops of this department were the infantry and artillery permanently attached to the line of forts which surrounded the capital. The only available field troops were Abercrombie's and Crawford's divisions of infantry, and Stahel's division of cavalry, about 13,000 in all, which formed a covering force to protect the capital from raiding parties. Abercrombie's division was composed of Stannard's Vermont brigade, Hay's New York brigade, and Fessenden's Maine brigade; Crawford's division consisted of three brigades of Pennsylvania troops; Stahel's cavalry division was made up of three brigades.

The *Middle Department* comprised the states of Maryland, Pennsylvania and West Virginia. Its troops, which formed the VIII corps, were widely scattered and had different duties to perform. Kelley's division of six brigades, about 12,000 men, was charged with the immediate protection of the Baltimore and Ohio railroad from Harper's Ferry to Grafton. For this purpose two brigades were strongly intrenched at Harper's Ferry, one was posted at Martinsburg, one at Romney, one at Cumberland and one at Grafton. Milroy's division, of 6,000 men, was at Winchester. It consisted of three brigades and was pushed to the front and intrenched to observe the Confederates in the Shenandoah Valley; Scammon's division of two brigades, 4,000 men, was at Charleston, West Virginia, on the Kanawha river, where it protected that state from invasion and also threatened invasion into southwestern Virginia. Lockwood's independent brigade was a field force of infantry and cavalry encamped near Baltimore; Morris' independent brigade was composed of the troops in the forts protecting the city of Baltimore; Brigg's independent brigade constituted the permanent city force of Baltimore; Averell's independent brigade guarded the Baltimore and Ohio railroad from Grafton to the Ohio river.

The *Department of Virginia* comprised that portion of southeastern Virginia which was held by Union troops. Its troops belonged to the IV, VII and IX corps. The IV army corps had one division, of about 5,000 men, under General Gordon at West Point on the York river and the other, of about the same strength, under the corps commander, General Keyes, at Yorktown. One division of the VII corps and one of the IX corps, about 18,000 men, the whole under General Peck, were at Suffolk, Virginia, south of the James, where they had recently successfully frustrated the attempt of General Longstreet to capture that place.

The *Department of North Carolina* comprised the coastal territory of that state occupied by Union troops. The main body of the XVIII corps, which occupied this department, was at Newberne, N. C.

The *Department of the South* comprised the coastal territory of South Carolina, Georgia and Florida occupied by Union troops. The X corps, which occupied this department, was distributed by divisions in the coastal islands between Charleston and the mouth of the Savannah river.

COMPARISON.—If the situation is viewed as a whole, it is apparent, that while the Union troops had the advantage of numbers, the Confederates were better disposed for concerted action and mutual support.

THE SOUTH.—The railroad running south from Fredericksburg to Savannah, via Richmond, Wilmington and Charleston, and the railroads running east and west from Richmond, permitted the Southern troops to be concentrated at any point in the theater of war which might be momentarily threatened. These roads enabled the authorities to withdraw troops from any department in which active operations were temporarily suspended and send them to the Army of Northern Virginia.

The supreme direction of military affairs was vested in the President of the Confederate States, Mr. Davis, a soldier by training and experience. Under him were only three independent commanders, General Lee, General Beauregard, and Major General Sam Jones, all educated and experienced soldiers. General Lee, who commanded not only the Army of Northern Virginia but also all the troops assigned to the defense of the capital, was fortunate in possessing the complete confidence of the civil authorities, the army, and the people of the Confederate states. This confidence he had well earned by the able manner in which he had conducted his military operations since June, 1862, when he first assumed command. It was the President who insisted on his assuming and retaining command of the departments of Richmond and of North Carolina, and he would have added the department of

South Carolina had General Lee been willing to accept it. Mr. Davis appreciated the disadvantages of divided control. General Lee, however, was unwilling to assume the responsibility for the conduct of affairs in departments to which he could not devote his entire attention. A compromise was therefore arrived at by which General Lee was consulted about all operations in the East which could even remotely affect him, and the operations of the Army of Northern Virginia were made of primary importance in the conduct of military affairs. On May 31st, 1863, General Lee was in command of over 106,000 effectives, or about seventy-eight per cent. of the Confederate force in the east, and had 74,000, or fifty-four per cent., in the Army of Northern Virginia and in Imboden's command. This was increased to 80,000, or fifty-nine per cent., before the opening of the campaign.

THE NORTH.—The sea, with its limited means of transportation, was the only route by which troops could be sent to, or withdrawn from, the Union departments of Virginia, North Carolina, and the South. These troops, could not, therefore, be quickly concentrated to take advantage of any weakness in the Confederate position, such as that created by the withdrawal of the Army of Northern Virginia from the front of Richmond, nor could they be quickly sent to reinforce the Army of the Potomac. In the general scheme of operations these forces merely assisted in the blockade of the coast and deprived the Confederate field armies of the troops retained by the coastal states for their own defense.

In the North, neither the President nor the Secretary of War were soldiers by education. Their principal adviser was General Halleck, the commander in chief of all the armies, who had had but little experience in conducting military operations and had not the genius of General Lee.

There were too many independent commands in the same territory. This deprived the Army of the Potomac of many men who should have been in its ranks. In the three departments commanded by Generals Schenck, Heintzleman, and Dix, there were more men than in the Army of the Potomac itself: at least one-half of these men would have been far more serviceable if united with that Army.

In the middle department, the troops of General Milroy at Winchester and those of General Kelley at Harper's Ferry were of no particular value. They were unable to prevent invasion of Maryland, as will be seen later, nor could they effectually protect the Baltimore and Ohio railroad from raiding parties.

In the Department of Washington, a covering force was unnecessary since by his instructions, the commander of the Army of the Potomac was to perform this duty. During the campaign most of these troops were incorporated with the Army of the Potomac.

While it was necessary to hold Fortress Monroe and Norfolk, the whole of General Dix's command was not necessary for this purpose. As *auxiliary* lines of operation, neither the line from Fortress Monroe to White House, nor that from Norfolk through Suffolk proved of any value during the Civil War.

Of the total number of troops in the East at this time, only forty-three per cent. were in the Army of the Potomac.

CHAPTER II.

CONFEDERATE PLANS.

An offensive movement by the Army of Northern Virginia was at this time almost forced on the Confederate authorities.

On the Mississippi river, General Pemberton was invested in Vicksburg by a superior force under General Grant, and it was apparent to those in authority that it was only a question of weeks when Pemberton must surrender.

At Port Hudson, a short distance south of Vicksburg, General Gardner was similarly invested by General Banks.

When these two places fell, the banks of the Mississippi, from the Ohio to the Gulf, would fall into the possession of the Union troops; Arkansas, Texas and Louisiana would then be separated from the Confederate states east of that river. This would be a serious blow to the Confederate cause both at home and abroad, and a brilliant counter-stroke at some other point in the theater of war was desirable to counteract its effects.

The forces in the Departments of Tennessee and of South Carolina had been so depleted in the attempt to form a relieving army for Vicksburg, that neither General Bragg in Tennessee, nor General Beauregard in South Carolina could be expected to make this counter-stroke. General D. H. Hill in North Carolina was also reduced to the defensive because of the forces drawn from his department to reinforce the

Army of Northern Virginia and cover Richmond from an attack along the north bank of the James.

The task of making this counter-stroke therefore fell on the Army of Northern Virginia, which had just been thoroughly reorganized and recruited and was in the highest state of efficiency.

There were still other reasons for an offensive movement.

The Confederate authorities found themselves short of food supplies and had great difficulty in filling the requisitions of the armies. No relief from this situation could be expected until the coming harvest, except by an invasion of the rich farm lands of Maryland and Pennsylvania.

The numerical strength of the Army of the Potomac was decreasing daily, due to the mustering out of the two-years men from New York who had volunteered in the spring of 1861, and the nine-months' militia from Pennsylvania and New Jersey who entered the service under the call of August, 1862. This army reached its minimum strength on June 20th, before the troops of other departments were united with it; it then numbered 89,000 men present for duty.

The Union troops in southeastern Virginia were giving the Confederate authorities some uneasiness about the safety of Richmond and it was thought that they would be drawn to Washington if that capital were threatened.

That General Lee was in favor of such a movement is shown by his correspondence. On June 8th he wrote to the Secretary of War as follows:

"As far as I can judge, there is nothing to be gained by this army remaining quietly on the defensive, which it must do unless it can be reinforced. I am aware that there is difficulty and hazard in taking the aggressive with so large an army in its front, intrenched behind a river, where it cannot be advantageously attacked. Unless it can be drawn out in a position to be assailed, it will take its own time to prepare and strengthen itself to renew its advance on Richmond, and force this army back within the intrenchments of that city. This may result in any event; still, I think it is worth a trial to prevent such a catastrophe. Still, if the Department thinks best to remain on the defensive, and guard, as far as possible, all the avenues of approach, and await the time of the enemy, I am ready to adopt this course. I think our southern coast might be held during the sickly season by local troops, aided by a small organized force, and the predatory excursions of the enemy be repressed. This would give us an active force in the field with which we might hope to make some impression on the enemy, both on our northern and western frontiers. Unless this can be done, I see little hope of accomplishing anything of importance. All our military preparations and organizations should be pressed forward with the greatest vigor, and every exertion made to obtain some material advantage in this campaign." (W. R. No. 45, p. 868.)

On June 25th, when his army was north of the Potomac river, in further development of this plan, he wrote to President Davis as follows:

"You will see that the apprehension for the safety of Washington and their own territory has aroused the Federal Government and the people to great exertions, and it is incumbent upon us to call forth all our energies. In addition to the 100,000 troops called for by President Lincoln to defend the frontier of Pennsylvania, you will see that he is concentrating other organized forces in Maryland. It is stated in the papers that they are all being withdrawn from Suffolk [Virginia], and according to General Buckner's report, Burnside and his corps are recalled from Kentucky.* It is reasonable to suppose that this would be the case if their apprehensions were once aroused.

"I think this should liberate the troops in the Carolinas, and enable Generals Buckner and Bragg to accomplish something in Ohio. It is plain that if all the Federal Army is concentrated upon this, it will result in our accomplishing nothing, and being compelled to return to Virginia. If the plan that I suggested the other day, of organizing an army, even in effigy, under General Beauregard at Culpeper Court House, can be carried into effect, much relief will be afforded. If even the brigades in Virginia and North Carolina, which Generals [D. H.] Hill and Elzey think cannot be spared, were ordered there at once, and General Beauregard were sent there, if he had to return to South Carolina, it would do more to protect both states from marauding expeditions of the enemy than anything else.

"I have not sufficient troops to maintain my communications, and, therefore, have to abandon them. I think I can throw General Hooker's army across the Potomac and draw troops from the south, embarrassing their plan of campaign in a measure, if I can do nothing more and have to return.

"I still hope that all things will end well for us at Vicksburg. At any rate, every effort should be made to bring about that result." (W. R. No. 45. p. 930.)

* One division of Burnside's corps was sent to Vicksburg at this time.

General Lee further explains the objects of his plan of campaign in the report he submitted at its close:

"The position occupied by the enemy opposite Fredricksburg being one in which he could not be attacked to advantage, it was determined to draw him from it. The execution of this project embraced the relief of the Shenandoah Valley from the troops that had occupied the lower part of it during the winter and spring, and, if practicable, the transfer of the centre of hostilities north of the Potomac. It was thought that the corresponding movements on the part of the enemy, to which those contemplated by us would probably give rise, might offer a fair opportunity to strike a blow at the army then commanded by General Hooker and that in any event that army would be compelled to leave Virginia and, possibly, to draw to its support troops designed to operate against other parts of the country. In this way it was supposed that the enemy's plan of campaign for the summer would be broken up and part of the season of active operations be consumed in the formation of new combinations and the preparations they would require. In addition to these advantages, it was hoped that other valuable results might be attained by military success. Actuated by these and other important considerations, that may hereafter be presented, the movement began June 3rd." (W. R. No. 44. p. 305.)

General Lee also recognized the fact that the South could hardly hope to succeed in a prolonged struggle with the united North. In a letter to President Davis, June 10, 1863, just before starting north, he writes:

"We should not therefore conceal from ourselves that our resources in men are constantly diminishing, and the disproportion in this respect between us and our enemies, if they continue united in their efforts to subjugate us, is constantly augmenting." (W. R. No. 45, p. 880.)

In this letter he advocates the encouragement of the peace party in the North by a discontinuance of speeches and newspaper articles in the South rejecting the terms of peace proposed by that party.

He hoped to make his invasion of the North another argument for those members of the peace party who declared that the South could never be subdued. In his circular order, directing that no supplies should be taken in the North except on regular requisition payable by the Confederate government, he also sought to avoid unnecessarily irritating the inhabitants of the invaded country, and to make the establishment of a permanent government in the South of material interest to them. (W. R. No. 45, p. 912.)

CHAPTER III.

THE OPPOSING ARMIES.

Army of Northern Virginia.—At the opening of the Gettysburg campaign, June 1, 1863, the Army of Northern Virginia had been under the command of General Robert E. Lee for one year. In that period it took part in the seven days' battles about Richmond, and in the campaigns and battles of Manassas, Antietam, Fredericksburg and Chancellorsville. The almost uniform success of its operations raised the morale of the entire army to the highest pitch and caused both officers and men to feel absolute confidence in their commander.

Notwithstanding the active operations of the year, the organization of the army had greatly changed since the army was hastily organized in front of Richmond. In the campaigns following the Peninsular campaign, Generals Longstreet and Jackson each commanded half the army; after Antietam, their commands were made *army corps* and both officers were promoted to the grade of lieutenant-general. Finding, however, that a corps of 30,000 men was too large to be conveniently handled, after the death of Jackson in May, 1863, General Lee formed his army into three corps. Upon his recommendation, the commands of the second and third corps were given to Major Generals Ewell and A. P. Hill, who were promoted to the grade of lieutenant-general. Both had commanded divisions in Jackson's corps, and the former was chosen by Jackson to succeed him.

Each of the three corps was composed of three divisions; a new division was made by uniting two of the six brigades of A. P. Hill's old division with the brigades of Pettigrew and Davis sent about May 31st from the departments of Richmond and North Carolina. Each of the nine divisions was composed of four brigades. All of the brigades had been reorganized so that, as far as possible, each was composed of troops from the same state.

Each brigade was formed of four or five regiments of ten companies each. On entering the service the strength of a regiment was about 1,000 men; as it had no depot battalion or company, its strength was rapidly diminished in active service.

Each division was commanded by a major general; each brigade by a brigadier general, if possible from the same state as his command.

The artillery of the army had, since the autumn of 1861, been under the control of Brigadier General William N. Pendleton. In front of Richmond the batteries, other than those of the reserve artillery of the army, were distributed among the brigades. From that period to Chancellorsville they were distributed among the divisions; in addition, in Longstreet's corps, there were two battalions of corps artillery. In the reorganization just before the Gettysburg campaign, the artillery reserve was broken up; one battalion of four batteries of four guns each was assigned to each division, and two similar battalions were assigned as corps artillery. The prescribed number of guns to each of the three corps was therefore eighty; this number was slightly exceeded, as the actual number with the army was two hundred and fifty-one. Of these twelve were 20-pounder siege rifles, two were Whitworth rifles, twenty-six were howitzers, and the remainder 3-inch rifles and 12-pounder smooth bore Napoleons in about equal numbers.

Although the batteries were attached to the divisions and army corps, the chief of artillery was responsible for the care and maintenance of his arm and supervised its employment on the battle-field. In this he was assisted by the chiefs of the corps artillery.

The cavalry of the army had, since the autumn of 1861, been under the command of Major General J. E. B. Stuart; it constituted an independent unit of the army. It had gradually grown from a brigade in the Peninsula campaign, to a division of three brigades in the Chancellorsville campaign. Just before the Gettysburg campaign General Lee still further strengthened it by adding to it the brigades of Generals A. G. Jenkins and W. E. Jones, which had been operating in the Shenandoah Valley, and that of General B. H. Robertson from North Carolina. To the cavalry division was attached a battalion of horse artillery of 24 guns, one battery for each brigade.

The actual strength of the army was approximately as follows, only the officers, and the enlisted men under arms, or effectives, being counted:

Infantry, May 31	59,448	W. R. No. 40, p. 845
Infantry joining after May 31st—		
Pettigrew's brigade (four regiments)	2,948	W. R. No. 26, p. 1086
Davis' brigade	2,414	W. R. No. 26, p. 1086
25th and 31st Va.	700	W. R. No. 40, pp. 714, 716
Total	65,510	

Infantry detached between May 31st and July 1st—

13th and 58th Va., and 54th N. C., left by Early at Winchester (approx.)........	1,200	W. R. No. 44, p. 464
Corse's brigade, Pickett's division, left near Hanover Station, Va..................	1,462	W. R. No. 49, p. 765
Ewell's losses at Winchester..............	230	W. R. No. 44, p. 335
Total............................	2,892	
Total infantry at time of battle............	62,618	
Artillery with army corps (thirteen battalions reported)	1,702	W. R. No. 40, p. 846
Alexander's and Garnett's battalions (estimated)	724	
Total............................	5,426	
Early's loss at Winchester.................	22	
Total artillery at time of battle............	5,404	
Total guns at time of battle with army corps	251	W. R. No. 49. p. 637
Cavalry—The brigades of Fitz Lee, Chambliss, Hampton, Jones and Robertson, May 31st	10,292	W. R. No. 40, p. 846
Jenkins' brigade	1,600	W. R. No. 44, p. 547
Total............................	11,892	
Detached and watching Harper's Ferry during invasion 12th Va. Cav.............	400	(estimated)
	11,492	
Casualties at Brandy Station, killed, wounded and missing.......................	485	
Same in Loudoun Valley.................	510	
Same at Hanover, etc....................	117	
Total............................	1,112	
Total Stuart's Cavalry Division north of Potomac on day of battle..............	10,380	
Imboden's command	2,100	Battles and Leaders, Vol. 3, p. 422.
Maryland Line, found by Ewell near Winchester and taken with him, about.....	600	W. R. No. 40, p. 602

TOTAL EFFECTIVES, INCLUDING OFFICERS.

Army of Northern Virginia—	Infantry	62,618
	Artillery	5,404
	Cavalry	10,380
		78,402
Maryland Line incorporated in Ewell's corps		600
		79,002
Imboden's command		2,100
		81,102

All the organizations above given were on the field of Gettysburg on July 3, and, with the exception of the cavalry brigades of Jones and Robertson and Imboden's command, (5,000 effectives), took part in the battle on one of the three days.

Before reaching the field the Confederate army lost an unknown number of men by sickness, straggling and desertion. This number cannot be estimated even approximately.

Army of the Potomac.—The Army of the Potomac, as constituted in the Gettysburg campaign, was composed mainly of the forces which had operated with General McClellan on the Peninsula, and General Pope in Northern Virginia. The two armies were united in September, 1862, and together took part in the Antietam campaign under General McClellan, in the Fredericksburg campaign under General Burnside, and in the Chancellorsville campaign under General Hooker. The battle of Gettysburg was fought under still another commander, General Meade.

Until General Hooker took command in January, 1863, the general organization of the army remained as established by General McClellan in the Peninsula campaign. The army was organized into army corps and an artillery reserve; in the later campaigns two corps were united to form a *grand division.* The enlisted strength, present for duty, of the army corps, averaged about 12,000 men, which was about two-thirds of its enrolled strength. Of the seven corps, which now formed the army, the II, III, V, and VI corps, had been with McClellan on the Peninsula; the I, XI and XII corps had formed Pope's Army of Virginia.

General Hooker abolished the grand divisions and reorganized both the artillery and cavalry.

The division consisted wholly of infantry; when it was separated from its corps, one or more batteries of corps artillery were assigned to it. The brigade and regimental organizations were as in the Confederate service.

Until the latter part of May, 1863, the batteries of artillery assigned to the army corps were generally distributed as divisional artillery; few of the corps had corps artillery. After the battle of Chancellorsville, the divisional artillery was abolished and corps artillery substituted for it. To each corps was assigned a brigade of artillery. The normal strength of a brigade was five batteries of six guns each, or thirty guns to each corps. The artillery reserve of the army was organized and commanded by Brigadier General H. J. Hunt, who in the Antietam campaign became Chief of the Artillery of the army. The artillery reserve was composed of five brigades of four or five batteries each, and had one hundred and fourteen guns. The actual number of guns with the army, not counting those with the cavalry, was three hundred and twenty. Of the guns of the corps artillery, one-half were 3-inch rifles and one-half were 12-pounder Napoleon smooth bores. Of the guns of the artillery reserve, eight were $4\frac{1}{2}$-inch siege rifles, which, however, did not reach the battlefield of Gettysburg, two were howitzers, and the remainder were 3-inch rifles and 12-pounder Napoleons in equal numbers.

In the Fredericksburg campaign, the cavalry was distributed among the grand divisions, one or two brigades to each grand division. Before the battle of Chancellorsville, General Hooker united all the cavalry into a corps of three divisions of two or three brigades each. General Stoneman was assigned to command the cavalry corps. After Chancellorsville, he was succeeded by General Pleasanton. It was reorganized by General Pleasanton into two divisions of three brigades each. A few days before the battle of Gettysburg, Stahel's division of cavalry was transferred from the Department of Washington to the Army of the Potomac and became the third division of the cavalry corps. General Kilpatrick was placed in command of this division, which was reorganized into two brigades and Generals Custer and Farnsworth were promoted to command them. To the cavalry corps were attached two brigades of artillery, each having five horse batteries of six rifled 3-inch guns.

On May 31, at the opening of the campaign, the strength of the seven army corps, the cavalry corps and the artillery reserve was. (W. R. No. 40, p. 574.)

	Present for Duty.
Seven army corps	86,690
Cavalry corps	10,192
Artillery reserve	3,898
Total	100,780

On June 30th the strength of the Army was (W. R. No. 43, p. 151):

Present for Duty Equipped.

Infantry	77,208	
Cavalry corps	14,973	present for duty, number equipped not stated.
Artillery reserve	3,898	
Cavalry with army corps	258	(estimated)
Stannard's Vermont brig	2,000	joined on evening of July 1st.)
Lockwood's Maryland brig	1,500	(estimated) joined July 2d.
	99,837	

Of the infantry all the organizations were on the field or in its immediate vicinity on July 3. Of the cavalry not more than five of the eight brigades were on the field at any one time. All of the artillery was on the field July 3, except two batteries of siege artillery.

Before reaching the field, the Union Army also lost an unknown number through sickness and straggling. This number, although it cannot be even approximately estimated, was probably large, as the army marched rapidly for several days before the battle in rainy weather and in the heat of the latter days of June. As the march was in a friendly country, the Union stragglers were in no danger of capture as were those of the Confederate army.

Although the aggregate strength of the seven corps had not changed materially between May 31 and June 30, the composition and organization of some of the corps had greatly changed, due to the expiration of the terms of service of many of the regiments. During the progress of the campaign, the deficiencies created by the withdrawal of these troops were made up by reinforcements from the Department of Washington and from the Middle Department. Of these reinforcements, Stannard's Vermont brigade was attached to the I corps; Hay's or Willard's brigade was attached to the II corps; Crawford's division of two brigades was attached to the V corps; and Lockwood's brigade was attached to the XII corps.

The cavalry corps was reinforced by the addition of Stahel's cavalry, a few days before the battle.

The artillery reserve was decreased during this period.

The aggregate strength of the infantry of the Union army present for duty, according to the returns, exceeded that of the corresponding units of the Confederate army, by about 15,000 men, which was about the strength of the infantry of the VI corps.

COMPARISON.—If the history and organization of the two armies is studied, it will appear that the Confederate army was the superior in all respects save numbers and equipment.

General Lee had commanded an army in five separate campaigns, and in more than twice as many important battles. General Hooker had commanded in a single campaign and battle, and General Meade was to be in supreme command only three days before he was to fight his first battle.

General Lee had the absolute confidence of his superiors and was free to operate as he saw fit. Both officers and men of his army were confident of success under his guidance, and he had confidence in the ability and hearty support of every officer and man of his army. Nearly all the general officers had been promoted on his recommendation.

General Hooker, having been unsuccessful in his one campaign, had not the confidence of his superiors, and therefore his suggestions and plans did not, as a rule, meet with cordial approval. His unsuccessful campaign of Chancellorsville must have caused both officers and men to distrust his ability to maneuver and fight his army.

While General Meade was given all the support possible by his superiors, his lack of experience made him hesitate to utilize his authority to its fullest extent. It led him to trust too much in the judgment of the corps commanders. He had not complete confidence in the ability of some of his principal assistants, and, because of his lack of experience and confidence in his own judgment, he did not create in all his subordinates confidence in himself.

The corps and division commanders of both armies were, as a rule, officers who had attained these commands only after successful service as brigade commanders; in the Confederate army all except one, General Rodes, were graduates of the U. S. Military Academy and had served in the regular army; in the Union army one of the corps and about one-third of the division commanders had entered the service from civil life as colonels of volunteers. Nearly all the older graduates had seen service in the Mexican war.

The ratio of officers to men was greater in the Confederate army than in the Union; according to the last returns before the battle the ratio of officers to men present for duty was .089 in the Confederate army, and .068 in the Union army. This ratio is an important factor in the value of an army when the officers have had a year's experience in the field, as had most of the officers of both armies.

The Confederate army corps and divisions were larger and more important units than those in the Union army. Their commanders had therefore greater power and responsibility. This was emphasized by giving the corps commanders the grade of lieutenant-general, and the division commanders that of major-general. In the Union army a major-general commanded the entire army, and officers of the same rank commanded corps and divisions; brigadier-generals commanded divisions and brigades. The clear lines of demarkation in the Confederate commanders undoubtedly made it easier for all commanders to control their subordinates.

The organization of the Confederate artillery was almost ideal. The chief of artillery had absolute control of its equipment and employment on the field, while it was so divided that each corps and division had its own unit. Its ammunition and equipment were much inferior to those of the artillery of the Union army. In the Union army the duties of the chief of artillery were not definitely fixed. The only guns in his absolute control were those of the artillery reserve. While at times extremely valuable, as a rule its guns were not so available as those of the corps artillery.

The cavalry of the Army of Northern Virginia had been, as an independent unit, under the personal charge of General J. E. B. Stuart from the beginning of the war. In the Union army it was not united into a single unit until February, 1863. It was still inexperienced as an independent unit, and it will be observed that General Hooker always gave it an infantry support when he sent it against General Stuart's cavalry.

The Confederate army had in the Gettysburg campaign fewer raw troops and fewer newly organized brigades and divisions than its adversary; this was a distinct advantage.

If the differences of the two armies are fairly weighed

it is not too much to say that the chances of success in the campaign about to be opened were in favor of General Lee, notwithstanding his numerical inferiority.

CHAPTER IV.

OPERATIONS FROM JUNE 1ST TO THE MORNING OF JULY 1ST.

MOVEMENT OF LONGSTREET AND EWELL TO CULPEPER.

On June 1, 1863, the main body of the Army of Northern Virginia lay on the south bank of the Rappahannock river, with headquarters at Fredericksburg. Stuart's cavalry division was at Culpeper, where five brigades were assembled. The divisions of Hood and Pickett, of Longstreet's corps, which had recently returned from southern Virginia, were not with the main army; Hood was at Verdiersville, southeast of Culpeper, and Pickett was near Hanover Junction with Pettigrew's brigade of Heth's division.

The main body of the Union army lay on the north bank of the Rappahannock and along the railroad to Acquia Creek with headquarters at Falmouth. The cavalry corps was assembled at Bealeton, on the Orange and Alexandria railroad, north of Culpeper.

On June 2 General Lee was informed that the Union troops had evacuated West Point on the York river and no longer threatened to advance on Richmond. He at once ordered the concentration of Ewell's corps and the divisions of Hood and McLaws of Longstreet's corps, at Culpeper. Hill's corps was left to hold Fredericksburg, and Pickett was ordered to join the army as soon as he was satisfied that his presence near Hanover Junction was no longer required.

On June 7 General Lee, having reached Culpeper in person, issued orders to the cavalry forces of Generals Jenkins and Imboden to prepare the way for his advance into the Shenandoah Valley.

General Jenkins was ordered to move down the Shenandoah valley at least as far as Strasburg and Front Royal, and there await the arrival of the infantry. In the meantime he was to obtain all possible information as to strength and position of the different Union posts in the lower valley. (W. R. No. 45, p. 865.)

General Imboden was to move from the upper Shenandoah valley into the South Potomac valley and engage the attention of the forces in and around Romney. He was to collect cattle for the main army and, if possible, by destroying bridges, to interrupt railway communication between Cumberland and Martinsburg. (W. R. 45, p. 865.)

While awaiting the execution of his preliminary orders, General Lee reviewed Stuart's cavalry corps on the plains about Brandy Station.

BATTLE OF BRANDY STATION OR FLEETWOOD.

The withdrawal of Ewell's forces from the Fredericksburg lines and their movement westward having been reported by the Union pickets, General Hooker ordered the cavalry corps to make a reconnaissance in the direction of Culpeper where it was reported that Stuart had united his cavalry. Two brigades, each composed of 1,500 infantry and a battery of horse artillery, were to accompany the cavalry.

Stuart, who was preparing for an offensive movement instead of anticipating an attack on himself, had only one brigade along the river. Jones' brigade was picketing the river about Beverly Ford. Of the other brigades Fitz Lee was further up the river, and W. H. F. Lee, Hampton and Robertson were near Brandy Station.

Early in the morning of June 9, Buford with his own division, the reserve brigade, and one brigade of infantry moved to Beverly Ford. It was a foggy morning and Jones' pickets were taken by surprise. The Confederates were being driven slowly back when Stuart came to Jones' assistance with W. H. F. Lee, Hampton and Fitz Lee. He stopped Buford's advance but could not drive him back as Buford was supported by infantry well posted.

In the meantime Gregg with two divisions and a brigade of infantry crossed without opposition at Kelly's Ford. He sent the third division on the direct road towards Culpeper; with his own division he marched towards Brandy Station; the infantry he sent northward on a road nearer the river than the one followed by the cavalry. He found Robertson alone at Brandy Station and drove him out. Stuart now came to Robertson's assistance with W. H. F. Lee and Jones. After desperate fighting with charge and counter charge, Gregg was obliged to fall back on his infantry. As Buford had been slowly pushing the cavalry in his front and advancing on the station, Gregg now found himself on Buford's left flank. Together they checked Stuart's advance until 5 P. M., when they were ordered to withdraw across the river. The third division of the Union cavalry corps met with a single Confederate regiment on the Culpeper road, which was easily brushed aside; it however took no further part in the day's operations and was withdrawn to the river with the other divisions.

The forces engaged on this day were about 10,000 Confederate cavalry, and 10,000 Union cavalry and infantry.

The reconnaissance had been successful in developing the strength of the Confederate cavalry and had also shown that they were supported by infantry. Rodes' division of Ewell's corps was sent by General Lee to Stuart's assistance and arrived at Brandy Station just at the close of the engagement, but too late to take an active part in it.

In the cavalry fight at Brandy Station General W. H. F Lee was severely wounded and Colonel Chambliss succeeded to the command of his brigade.

CAPTURE OF WINCHESTER.

On the following day, June 10, before the Union army could profit by the knowledge obtained and frustrate his movement, Ewell was ordered to march on Winchester via Front Royal and there, if possible, capture the division of Major General Milroy.

Winchester, in the Shenandoah valley, was not occupied by Union forces after its evacuation in the Antietam campaign until January, 1863, when Brigadier General Milroy was sent there. Although General Halleck had given repeated instructions that this town was not to be occupied in force and its troops were to be withdrawn promptly whenever threatened by superior numbers, nevertheless, it was gradually converted into a base for operations in the Shenandoah valley, by the formation of a weak entrenched camp and by the increase of its garrison to a division of three brigades (W. R. No. 31, p. 947; No. 40, pps. 132, 317, 453.)

Finally on June 10, when he learned of the concentration of the Confederate forces at Culpeper, General Halleck ordered the immediate evacuation of Winchester. This order was sent to General Milroy by General Schenck, chief of staff, but afterwards, on General Milroy's representations, was in effect rescinded by General Schenck; in view, however, of an expected attack, the brigade which had occupied Berryville was ordered to Winchester. (W. R. No. 45, p. 50.)

Ewell reached Cedarville, just beyond Front Royal, on the evening of June 12 and was there joined by Jenkins' cavalry.

From the information given him, he decided to send Rodes' division with Jenkins' cavalry, on the 13th, to capture the brigade at Berryville, and then have Rodes push on and capture the post at Martinsburg. Early's division was to move to Newtown and thence to Winchester on the Strasburg road, while Johnson's division was to move on the direct road from Cedarville.

The force under General Rodes moved on Berryville and while making dispositions to surround it, found that the Union forces had retreated, leaving only a small rear guard. Rodes followed the retreating force to Summit Point on the railroad, where the Union brigade had turned off towards Winchester; he there bivouacked for the night after a march of twenty miles. He attempted to report to General Ewell for further orders, but was unable to do so; therefore, in pursuance of his original orders, on the morning of June 14 he marched on Martinsburg and reached that town late in the afternoon after a long march. The garrison of that post having orders to retreat if menaced by superior forces, made its escape during the night. The cavalry was pursued by Jenkins as far as Williamsport, but the infantry escaped unobserved to Shepherdstown.

Generals Early and Johnson, marching on Winchester on the 13th, met the main force under General Milroy north of Kernstown and from that point drove him to his fortified positions just south of Winchester. On the morning of the 14th, finding the positions south of Winchester too strong to warrant an assault, General Early was directed to make a retrograde and turning movement with three of his brigades and assault the fortifications north and west of the town. To conceal his design, Gordon's brigade was left south of Winchester and Johnson's division threatened the lines east of the town. Early reached his point of attack without being discovered, and having prepared the way with twenty guns carried the works in his front by assault. It was now dark and too late to attack the other works in the line of defense.

That night Milroy decided to retreat on Harper's Ferry. This movement was anticipated by General Ewell, who sent three brigades of Johnson's division to prevent his escape. Johnson reached the Martinsburg road at a point five miles north-east of the town, with two of his brigades, some time after midnight. Just at that moment the head of the retreating troops reached the same point. The opposing troops fought in the early dawn and the Confederates were about to be overwhelmed when the opportune arrival of Johnson's third brigade, which had been accidentally left behind, turned the tide. The Union retreat then became a rout, in which about 4,000 men, 23 guns, and 300 wagons were captured. The remainder of Milroy's force reached the Potomac either at Harper's Ferry or at Hancock. Rodes' division was too far from the lines of retreat to intercept either of the columns.

Leaving three of Early's regiments to guard the prisoners, Ewell moved to the Potomac, where Rodes crossed at Williamsport on the 15th, and on the 19th moved to Hagerstown; on the 18th Johnson crossed at Sharpsburg, and on the same day Early moved to Shepherdstown. In these positions they remained until the arrival of Hill's corps.

ACTION TAKEN BY THE PRESIDENT TO MEET THE INVASION OF NORTHERN STATES.

In view of a probable invasion of Pennsylvania, on June 10th, President Lincoln formed two new departments: that of the Susquehanna, with headquarters at Harrisburg, under General Couch, then commanding the II corps; and that of the Monongahela with headquarters at Pittsburg under General W. F. H. Brooks. (W. R. No. 45, p. 55.)

On June 15, when the news of the disaster at Winchester was received in Washington, the President issued a call for 100,000 militia to be furnished by the threatened states, Pennsylvania, Maryland, West Virginia and Ohio, and had letters written to the Governor of New York asking for 20,000 men, and to the governors of the other states asking for such regiments as could be furnished at once. (W. R. No. 45, p. 136.)

On June 24th, the Middle Department was divided into two parts: the part west of the Shenandoah Valley became the Department of West Virginia, under General B. F. Kelley. (W. R. No. 45, p. 299.)

RETROGRADE MOVEMENT OF THE ARMY OF THE POTOMAC.

As soon as General Hooker became aware of the flank movement of the Confederate army, he wrote to the President suggesting that he be allowed to attack the force left at Fredericksburg, should it develop that Lee was dividing his army to make an offensive movement. He was advised by both the President and General Halleck not to move across the Rappahannock. Again, on June 10th, he requested permission to move on Richmond. This was also refused and he was directed to make his movements conform to those of General Lee, attacking him whenever possible. (W. R. No. 43, pp. 31-35.)

On the 11th of June therefore he began to withdraw from the Rappahannock and concentrate along the Orange and Alexandria Railroad near Manassas Junction, en route for the Potomac near Leesburg.

On June 15, the I, III, V and XI corps were already at Manassas Junction. The other corps were on their right and rear marching on Fairfax Court House. The cavalry corps had remained at and near Warrenton to cover this movement, and on the 15th and 16th joined the other corps about Manassas Junction. On the 17th, the V corps was ordered to Gum Springs, near the Little River turnpike northwest of

Manassas; to cover this movement Gregg's division of cavalry was ordered to Aldie.

CONCENTRATION OF THE CONFEDERATE ARMY NEAR THE POTOMAC.

When General Lee learned of the withdrawal of Hooker from the Rappahannock, he ordered Hill to move to Culpeper and thence to Winchester, and Longstreet to move along the east slope of the Blue Ridge mountains from Culpeper northward to Ashby and Snicker gaps. The exposed flanks of Longstreet and Hill were to be covered by Stuart's cavalry.

Hill, being entirely protected and having free roads, left Fredericksburg on the 14th of June and reached the Potomac, at Shepherdstown, June 23.

Longstreet moved northward on the 16th, as soon as Hill's leading division was at hand, and on the 19th he halted with Hood's division in Snicker Gap, McLaw's division in Ashby Gap, and Pickett's in the mountain passes between the other two divisions.

In this movement Fitz Lee's cavalry brigade formed his advance guard as far as the Manassas Gap railroad and the brigades of Chambliss and Robertson formed his flank guard. Hampton was left on the Rappahannock to protect Hill, and Jones was between Hampton and Robertson.

CAVALRY ENGAGEMENTS IN LOUDOUN VALLEY.

On the 17th of June, after escorting Longstreet's corps to the passes of the Blue Ridge mountains, Fitz Lee's cavalry brigade moved eastward towards Aldie Gap in the Bull Run mountains. On the same day Gregg's cavalry division, after escorting the V corps to Gum Springs, moved westward through Aldie Gap to ascertain the position of the enemy. The two forces met just west of Aldie, where the brigades of Fitz Lee and Kilpatrick had a spirited encounter. On the approach of Gregg's remaining brigades, Lee retired.

On the following day the brigades of Chambliss and Robertson having arrived, they were posted on the road leading from Aldie to Ashby Gap and about six miles from the former place. Fitz Lee was posted to their left and rear. On the 19th, when Jones' brigade arrived, it relieved Fitz Lee's brigade, which was then moved to the road from Aldie to Snicker Gap. On the 20th, Hampton's brigade arrived and was attached to the right wing.

There was some skirmishing between the two cavalry commands at Middleburg June 18 and 19 on the Ashby Gap road, but no serious movement was made until June 21. On June 21 Pleasanton was authorized to attack the cavalry in his front and to ascertain what Confederate force was east of the Blue Ridge mountains. In this attack he was to be supported by the V corps, which had been ordered to Aldie. On the 21st, therefore, with Gregg's cavalry division and Barnes' division of the V corps on the Ashby Gap road, and with Buford's cavalry division on his right, Pleasanton moved against Stuart and drove him from his intrenched position at Upperville into the Blue Ridge passes. In the defense of these passes Stuart's command was temporarily relieved by Longstreet's infantry. Only Vincent's brigade of the V corps was actively engaged in this movement.

On the 22nd, Pleasanton's command returned to Aldie, where he remained with the V corps until the 26th, when he moved to Leesburg. Stuart's cavalry returned to the valley between the Bull Run and the Blue Ridge mountains, where he remained until the 25th.

CONFEDERATE ADVANCE FROM THE POTOMAC.

As it was apparent, on the 22nd of June, that the Union army had no intention of advancing across the Loudoun Valley, General Lee issued the orders for the advance of his army into Pennsylvania.

On June 23 Ewell's corps moved northward. His orders were to move to the Susquehanna, collecting supplies on the way, and to seize Harrisburg if he could. (W. R. No. 45, p. 914.) Ewell's main body, consisting of most of Jenkins' cavalry and the infantry divisions of Rodes and Johnson, marched through Hagerstown, Greencastle and Chambersburg, to Carlisle, which they reached June 27. His left flank was protected by Steuart's infantry brigade with the Maryland Line, which moved from Greencastle to Chambersburg, via McConnellsburg, and his right flank by Early's division with one of Jenkins' regiments, the 17th Va. cavalry, which moved through Boonsborough and Waynesborough to Greenwood on the Chambersburg-Gettysburg road east of Fayetteville. The march was a slow one, because of the flanking columns and because of the collection of supplies in a hostile country. From Carlisle, Jenkins' cavalry was sent to reconnoitre the defenses of Harrisburg, and on the morning of June 29 Ewell was preparing to move against it with his infantry, when he received orders to move towards Gettysburg.

Early's division, on reaching Greenwood, turned eastward through the South mountains. Gordon's brigade, with White's battalion of Jones' cavalry brigade, moved directly on Gettysburg, and the other brigades, with the 17th Va. regiment of Jenkins' cavalry, moved to Mummasburg. From Mummasburg Hays' brigade was sent to Gettysburg. The right wing of Early's division then moved on the turnpike to York, the cavalry going to Hanover Junction, while his left wing moved by the country roads through Hunterstown and Berlin. The division reached York on June 28 and Gordon was at once sent to seize the bridge over the Susquehanna at Columbia. In this he was unsuccessful, as the bridge was burned by the local troops. Early's division was encamped around York, collecting supplies and destroying the railroad, when on June 29 its commander received orders to join his corps near Cashtown.

Hill's corps crossed the Potomac on the 24th of June and marched through Hagerstown and Chambersburg to Fayetteville, where it arrived June 27. On the 29th, Heth's division was sent across the South mountains to Cashtown.

Longstreet's corps crossed the Potomac on the 25th and 26th of June and reached Chambersburg on the 27th. On the 29th, the divisions of McLaws and Hood were sent to Greenwood, just beyond Fayetteville; Law's brigade of

Hood's division was posted as an outpost at New Guilford on the road to Fairfield.

General Imboden, who had carried out his preliminary instructions and had destroyed many of the bridges, block-houses and water tanks along the Baltimore and Ohio railroad, was ordered to keep on the left flank of Ewell's corps and gather information and collect supplies. (W. R. No. 45, pp. 905 and 924.) He therefore assembled his force at Hancock on the Potomac, where General Lee's courier found him on June 29.

On General Stuart's cavalry, General Lee depended for information of the movements of the Army of the Potomac. He was therefore directed to cross the Potomac with three brigades as soon as their services could be spared from the defense of the Blue Ridge passes. He was directed to place himself on Ewell's right flank and was informed that one of Ewell's columns would probably march to Emmitsburg. (W. R. No. 45, p. 913.)

General Stuart, it appears, had proposed a plan by which he was to move through Hopewell Gap and reach the Potomac by moving around the rear and right flank of the Union army. This plan was referred to by General Longstreet in transmitting General Lee's order to General Stuart and in his letter to General Lee informing him of its transmittal. The movement met the approval of General Longstreet. (W. R. No. 45, p. 915.)

General Lee finally gave General Stuart permission to move by the route which seemed best to General Stuart. He, however, added that if the Union army was not moving northward, he advised him to move into the Shenandoah valley on the night of the 23rd, cross the Potomac at Sheperdstown and move toward Frederick, Maryland. (W. R. No. 45, p. 923.)

Acting under the discretionary orders thus given him, on June 24 General Stuart assembled the brigades of Fitz Lee, Chambliss and Hampton on the Manassas Gap railroad, a short distance west of Thoroughfare Gap. His original intention was to move through Hopewell Gap, north of Thoroughfare Gap, and thence through Gum Springs to Seneca Falls. However, at this time the Union forces were too far south to permit it. In Aldie Gap were the V and cavalry corps; in Thoroughfare Gap and Gainesville was the II corps; at Bristoe station, south of Manassas and Centerville, was the VI corps. Not knowing fully the position of the Union forces, Stuart went through a gap just south of Thoroughfare Gap early in the morning of June 25 and shortly ran into the II corps. The whole day was spent in reconnoitering the Union forces and in a useless engagement with them. On the 26th he marched by Brentsville to a point on the Occoquan River below its junction with the Bull Run. On the following day, June 27, he made a long march and that night crossed the Potomac at Rowser's ford. On June 28, finding that the entire Union army was at and near Frederick, he marched northward through Rockville to Brooksville. Near Rockville he captured a train of more than one hundred and twenty-five wagons, which he took with him. On the night of June 29, when the other commanders had received orders to march on Gettysburg, he had just reached Union Mills, half way between Westminster and Littlestown.

The cavalry brigades of Robertson and Jones were left

by Stuart to guard the passes in the Blue Ridge mountains as long as the Union army was in their front. When that army crossed the Potomac, General Robertson, who was in command, was to withdraw beyond the Shenandoah River, cross the Potomac and place himself on the right and rear of the Confederate army. One battalion of Jones' brigade had been attached to Ewell's corps and a regiment was on picket duty near Harper's Ferry. Robertson's brigade and the three remaining regiments of Jones' brigade were assembled at Berryville on June 29, preparatory to crossing the Potomac and joining the army, when they received orders to march to Gettysburg.

CONCENTRATION OF THE UNION ARMY AT FREDERICK, MARYLAND.

On June 17 the Union army began its movement northward from Manassas for the Potomac near Leesburg, Virginia, where bridges were to be constructed for its crossing. This movement was covered by the II corps at Thoroughfare Gap and the V and cavalry corps at Aldie Gap.

On the 25th of June the Union army began crossing the Potomac and moving towards Frederick. It was preceded by Stahel's cavalry division of the Department of Washington, which had been sent from Fairfax Court House for this purpose.

On June 26 the cavalry corps, which formed the rear guard, left Aldie and marched to Leesburg.

On June 28 the entire army was concentrated in a circular area whose diameter is twenty miles and whose center is Frederick. On the night of June 27-28, General Hooker was, at his own request, relieved from the command of the army. He stated that he submitted his resignation because the forces commanded by him were inadequate to meet his adversary and also cover Washington and Harper's Ferry as required by his instructions. He had just previously recommended the abandonment of Harper's Ferry and the reinforcement of his army by its large garrison. This recommendation had been disapproved. (W. R. No. 43, p. 60.)

General George G. Meade, commander of the V army corps, was appointed by President Lincoln to succeed General Hooker; he assumed command on the morning of June 28. His instructions were to maneuver and fight in such a way as to cover Washington and Baltimore. He was placed in command of all forces in his sphere of operations, including those of Harper's Ferry, and was authorized to remove from, or assign to, the command of any part of his army, any officer whom he saw fit. (W. R. No. 43, pp. 61, 62, 63.)

In pursuance of his orders and from definite information that the corps of Ewell, Hill and Longstreet had marched north of Hagerstown, he decided to move his army to the line Emmitsburg-Hanover, covering Baltimore and Washington. Westminster, Maryland, was made his new base.

On the 29th of June, Buford with two brigades of his cavalry division, marched from Middletown, west of Frederick, through Boonsborough towards Fairfield; his third brigade was sent to Mechanicstown. Gregg's cavalry division marched from Ridgeville, on Parr's Ridge east of Frederick, to New Windsor on the road to Westminster; Kilpatrick's cavalry division marched from Frederick to Littlestown, where it was joined by the 5th and 6th Michigan regiments, which were returning from a reconnaissance to Gettysburg, which they reached June 28.

On ~~the same day~~ June 29, the I and XI corps moved to Emmitsburg; the III to Taneytown; the II to Union~~ville~~town near Westminster; the V to Liberty in rear of the II; the XII to Bruceville in rear of the III, and the VI to New Windsor on the road to Westminster.

THE ORDER FOR THE CONCENTRATION OF THE CONFEDERATE ARMY AT GETTYSBURG.

On the night of June 28, General Lee learned from a spy that the Union army was around Frederick. This was his first intimation that the Army of the Potomac was north of the Potomac River. He had assumed that General Stuart would precede the Union army across the river and give him due notice of that fact. To prevent the Union commander from moving still further westward and intercepting his communications he decided to concentrate his army east of the South mountains, either at Cashtown or Gettysburg. Orders to this effect were therefore sent to General Ewell and General Robertson, and orders were sent to General Imboden to move to Chambersburg and relieve General Pickett of the charge of that town and the trains.

POSITIONS OF THE ARMIES ON THE NIGHT OF JUNE 29.

By the evening of June 29, these orders reached all the separate Confederate commanders, except Stuart. The military situation was then as follows: Of the nine infantry divisions of the Confederate army, one was at York; two at

Carlisle; one at Chambersburg; four near Fayetteville; and one at Cashtown. Stuart, with three cavalry brigades, was at Union Mills en route for Littlestown; Robertson, with two cavalry brigades, was at Berryville in the Shenandoah valley, and Imboden was at Hancock on the Potomac river.

Of the Union forces, Buford with two cavalry brigades was moving on Fairfield and Gettysburg; Gregg's cavalry division was moving on Westminster; and Kilpatrick's division was at Littlestown en route for Hanover. The infantry corps were assembled on the Emmitsburg-Taneytown-Westminster road.

POSITIONS OF THE ARMIES ON THE NIGHT OF JUNE 30.

The movements of each army on June 30 conformed to its general plan of operations.

The Confederate army corps moved towards Cashtown and Gettysburg, and the Union army corps deployed on the general line Emmitsburg-Hanover.

As Buford was advancing towards Fairfield he discovered that the town was in the possession of the Confederates. Not wishing his presence to be known, he turned eastward until he reached the Emmitsburg-Gettysburg road and upon that marched to Gettysburg, which he reached at 11 A.M. At the same hour Pettigrew's brigade of Heth's division had just reached the outskirts of the town on its march from Cashtown for supplies.

Before the two commands came into actual conflict Pettigrew withdrew to Cashtown and reported the presence of Union cavalry at Gettysburg. Buford took possession of the town and posted his pickets so as to watch all the roads from the Fairfield road on his left, to the York road on his right. Beyond the picket line, in all directions, he sent reconnoitering patrols.

Gamble's brigade, which was on the left of the line, was between the town and Willoughby Run and between the Fairfield and Cashtown roads. Its picket line was on a ridge just west of the creek and an advanced picket of the 8th Illinois cavalry was on a ridge nearly a mile and a half west of Willoughby Run. Devin's brigade was north of the town.

On the night of June 29-30, Stuart, who was at Union Mills, learned that the Union troops had just marched into Littlestown. On the morning of the 30th therefore, he marched to Hanover to avoid them. As his advance guard marched into that town it encountered Kilpatrick's rear guard which was just passing through it. Although outnumbering Kilpatrick, Stuart was hampered by the long wagon train he had captured and was desirous of avoiding an engagement which would still further delay him. He therefore turned eastward, made a forced march, and reached Dover on the morning of July 1.

On the night of the 30th, the positions of the opposing forces were as follows:

At Heidlersburg, north of Gettysburg, was General Ewell with Rodes' division, and three miles east of that town, was Early's division. At Cashtown were the divisions of Heth and Pender of Hill's corps. Across the mountains, in the vicinity of Fayetteville, were the divisions of Anderson, McLaws and Hood. Johnson's division was on the Carlisle-Chambersburg road about five miles from the latter place.

Pickett's division was still at Chambersburg. Stuart's cavalry was making a night march between Jefferson and Dover, Penn.; Robertson was near Martinsburg, Va.; and Imboden near McConnellsburg, Penn.

The main line of the Union army was composed of Merritt's cavalry brigade at Mechanicstown; the I corps at Marsh Run, midway between Emmitsburg and Gettysburg; the XI corps at Emmitsburg; the III corps a short distance east of Emmitsburg, on the Taneytown road; the XII corps at Littlestown; the V corps at Union Mills; and the VI corps near Manchester. The II corps was in the second line at Union~~ville~~town west of Westminster. Buford, with two cavalry brigades, was at Gettysburg; Kilpatrick's division was at Hanover; Gregg's cavalry division had two brigades at Westminster and one at New Windsor, Md.

CHAPTER V.

BATTLE OF JULY 1.

MOVEMENTS PRECEDING THE BATTLE.

It was not the intention of either of the army commanders to begin a decisive battle on this day.

CONFEDERATE ARMY.—Of his nine infantry divisions, General Lee had only four east of the South mountains, and these were not concentrated. All of the other divisions were still west of the mountains, echeloned along the single road all must use to pass this barrier. As Imboden had not arrived, Pickett was still guarding the army trains. Of his cavalry, General Lee had a single brigade with his army; of Stuart's position he had no knowledge whatever and Robertson was still far in the rear. Of the position of the Union army he knew only that Pettigrew had seen some cavalry entering Gettysburg. No new orders were therefore given for July 1; the corps commanders were allowed to complete the concentration and were informed that no general engagement was desired until the army was concentrated. General Lee remained with General Longstreet until noon, when Longstreet's corps began to cross the mountains.

The Confederate point of concentration having been shifted by General Hill from Cashtown to Gettysburg, on the evening of June 30 he ordered the division of Heth to be followed by that of Pender to move to Gettysburg, and that of Anderson to cross the mountains to Cashtown. He also sent

Ewell word that he was going through Cashtown to Gettysburg.

Early in the morning Heth started for Gettysburg, accompanied by one battalion of the corps artillery in place of his own. Pender, who was bivouacked some distance in his rear, did not move until an hour or two later, and was followed by the other battalion of corps artillery.

On the morning of July 1, General Ewell started with Rodes' division to march directly from Heidlersburg to Cashtown; before he reached Middletown he was met by General Hill's messenger and therefore moved from Middletown towards Gettysburg. General Early was ordered by General Ewell to move southward until he reached the Hunterstown-Mummasburg road and thence move west to Cashtown. He was moving on the Heidlersburg-Gettysburg road southward when he was overtaken by a message from General Ewell to move to Gettysburg. General Johnson, with Ewell's corps trains, was leaving the Carlisle-Chambersburg road about five miles north of Chambersburg and marching on a cross road for Fayetteville.

On the morning of July 1, General Longstreet was preparing to move across the mountains with the division of McLaws and Hood as soon as the road was clear; Pickett was to join them east of the mountains as soon as relieved by Imboden, who was expected that day.

UNION ARMY.—General Meade's orders for the movements on July 1 were issued before he knew definitely of the withdrawal of the Confederate forces from York and Carlisle. Although he took the precaution to have all the unnecessary trains sent to the rear and the army ready for battle, he did not intend that the movements ordered should involve any of his army corps in an engagement with the enemy. His orders were, the I corps to Gettysburg; the XI to supporting distance from the I corps; the III corps to Emmitsburg; the XII corps to Two Taverns, five miles from Gettysburg on the Littlestown road; the II corps to Taneytown; and the V corps to Hanover. The VI corps was to remain at Manchester.

After issuing these orders, he learned that Carlisle and York were being evacuated and that the Confederate army was being concentrated probably to make a sudden descent upon him. The point of concentration could not be accurately fixed, but he thought it would be somewhere on the road from Chambersburg to York, through Cashtown and Mummasburg.

As his army was deployed over a long line, it was necessary for him to concentrate for battle either in advance of his present position or in rear of it. Acting upon the hypothesis that Reynolds might be attacked by the enemy and compelled to fall back, he selected a line along Pipe Creek between Middleburg and Manchester, and sent a circular to his corps commanders to study the roads by which it would be necessary for them to move in order to concentrate there. (W. R. No. 45, p. 458.)

He also wrote to General Reynolds giving him an outline of the military situation and asking his views as to the advisability of concentrating the army in the vicinity of Gettysburg, in view of the information given him in General Meade's letter, and the knowledge General Reynolds had of the nature of the country, etc. He informed General Reynolds that the orders for July 1 were issued before the evacuation of Carlisle and York were known. (W R. No. 45, p. 460.)

Having great confidence in the ability and judgment of General Reynolds, he had placed him in command of the III and XI corps in addition to his own, with instructions "to make such dispositions and give such orders as circumstances may require." (W. R. No. 45, p. 414.)

General Reynolds ordered the XI corps to follow his own to Gettysburg; all the other corps moved according to the orders of the day.

At 8 A.M. on July 1, Wadsworth's division of the I corps, which was bivouacked on the Emmitsburg-Gettysburg road near the Marsh Creek bridge, with Hall's battery of artillery, moved towards Gettysburg. General Reynolds rode with it. The second division, which was bivouacked on a road running from the Marsh Creek bridge to Fairfield, and the third division, which was in rear of the first division, started about 9:30 A.M. The XI corps started from Emmitsburg some time after 8 A.M.; Barlow's division moved by the direct road, but the other divisions being encamped east of the town moved by a country road which entered the Gettysburg-Taneytown road about five miles south of Gettysburg.

Neither the Confederate nor the Union movements were made either with haste or with extreme caution; the rains of the preceding day had made the roads difficult and neither side was expecting to meet the enemy in force. General Hill assumed that Gettysburg was held by a small cavalry force, which could readily be brushed aside by Heth's division. General Reynolds assumed that no enemy could be met en route, since Buford held Gettysburg.

As no other troops received orders or requests to move to Gettysburg in time, the only troops engaged in the battle of July 1 were the Confederate infantry divisions of Heth, Pender, Rodes and Early, seventeen brigades, and the Union corps of Reynolds and Howard, assisted by Buford's cavalry, twelve infantry and two cavalry brigades.

BATTLEFIELD OF JULY 1.

The field upon which the opposing forces met on this day lies north and west of the town of Gettysburg and is limited on the east and west by two insignificant streams called *Rock Creek* and *Willoughby Run*. It is limited on the south by the *Hanover*, and the *Fairfield* or *Hagerstown road*, and is traversed by the roads running to *Cashtown*, *Mummasburg*, *Middletown*, *Carlisle*, *Heidlersburg* and *York*.

About a half mile west of the town is a ridge, running due north and south, which forms the divide between the two streams. At its extreme north end, where, in the reports of the battle it is called *Oak Hill* or *Oak Ridge*, it rises one hundred and twenty feet above the creeks; opposite the town, where it is called *Seminary Ridge*, it rises to a height of only sixty feet. Its eastern slope is quite steep; its western very gentle.

Between Rock Creek and this ridge is a level plain, rising only twenty feet above the creek. Its only elevation is *Barlow's Knoll*, near the Rock Creek crossing of the Heidlersburg road. This knoll rises about thirty feet above the level of the plain and is connected with Seminary Ridge by a low divide.

The field, between Seminary Ridge and Willoughby Run, is traversed by two intermediate ridges, both radiating from Oak Hill. The first ridge, west of Seminary Ridge, which may be called *Reynolds Ridge*, since it is now followed by *Reynolds Avenue*, crosses the Fairfield road a half mile west of Seminary Ridge. North of the Cashtown road, the depression between this and Seminary Ridge is hardly perceptible, but at the Fairfield road the bottom of the depression is fifteen feet below Reynolds Ridge, and thirty-five feet below Seminary Ridge.

Between Reynolds Ridge and Willoughby Run and about three hundred and fifty yards from the former is *McPherson Ridge*, now followed by *Stone* and *Buford Avenues*. This ridge is pierced by *Bender's Ravine* north of the Cashtown road and terminates in the *McPherson Woods*, about a quarter of a mile south of that road. The depression between this and Reynolds Ridge is a shallow one.

The *Maryland and Western Railroad*, at the time of the battle graded but not completed, runs nearly parallel to and about a hundred yards north of the Cashtown road. It traversed all the ridges above described in cuts four to six feet deep.

The field was generally open, but scattered over it were groves of small extent. The most notable of these were the *McPherson Woods*, a grove two hundred yards wide, extending from Willoughby Run to Reynolds Ridge; *Shead's Grove*, covering the west slope of Oak Ridge from the railroad northward; and the *Seminary Grove*, between the Cashtown and Fairfield roads.

In rear of the position and the town, near the junction

of the roads from Emmitsburg, Taneytown and Two Taverns, is *Cemetery Hill.* This elevation commands the plain north of it by about one hundred feet and the ridge at the Seminary building by about fifty feet. Since it commanded all the roads by which the I and XI corps could retreat or be reinforced, it was the true key-point of the position.

HETH'S ATTACK ON THE MORNING OF JULY 1.

On the morning of July 1, as Heth passed the Marsh Creek bridge on the Cashtown road, he encountered the advanced picket of the 8th Illinois, which fired the first shot of the battle. As the Confederates were seen advancing, by the main picket line on the ridge west of Willoughby Run, Gamble's cavalry brigade, which was south of the Cashtown road, was at once deployed along Reynolds and McPherson ridges from the railroad towards the Fairfield road. Calef's horse battery was posted along his line; two guns on the right of the Cashtown road, two on the left, all on McPherson Ridge, and two still further to the left, beyond McPherson's woods. Devin's brigade was north of Bender's Ravine. It was Buford's intention to hold Gettysburg until he was compelled to retire or ordered to do so. He knew that the I corps was on its way to Gettysburg and not far distant.

When General Heth encountered the advanced picket line near the Marsh Creek bridge, he deployed Archer's brigade, which was in advance, on the south side of the Cashtown road, and Davis' brigade, which followed it, on the north side. By this advance line the Union pickets were gradually driven back until the main Union line east of Willoughby Run was encountered. Then the artillery was ordered to the front to silence Calef's battery, and Archer and Davis prepared to attack the cavalry's line of battle.

It was probably nearly 11 A.M. when the two Confederate brigades waited near Willoughby Run for the artillery to silence Calef's guns before attacking the line on McPherson Ridge. At this time Wadsworth's division was approaching Gettysburg.

Upon hearing that Buford was attacked, General Reynolds rode at once to the field and ordered Wadsworth's division to march over Seminary Ridge and go to the support of the cavalry.

Davis' brigade was probably crossing the creek, when Wadsworth approached the field from the Seminary. Cutler's brigade, being in advance, was sent with Hall's battery to check the advance of the Confederates north of McPherson's Woods. Hall's battery came in position on the north side of the Cashtown road near Calef's right platoon and the brigade line was formed by placing two regiments on his left and three north of the railroad. Meredith's brigade of Wadsworth's division was sent against Archer's brigade, which was just emerging from McPherson's Woods as it arrived. Four of Meredith's regiments charged and drove Archer back into the creek where he with a number of his men were captured; the other regiment of Meredith's brigade was held in reserve. General Reynolds was killed in the McPherson Woods by one of Archer's men and the command of the corps fell to General Doubleday.

Davis was at first more successful than Archer; he compelled two regiments of the right wing of Cutler's brigade, not fully deployed, to fall back to Seminary Ridge and nearly captured Hall's battery before it could retire. However, as he was advancing over Reynolds Ridge, with two regiments north and one south of the railroad, he was attacked in flank by Meredith's reserve regiment and the two regiments of Cutler's left wing. His right regiment was driven into the railroad cut where a large part of it was captured. The other regiments of Davis' brigade thereupon retired across Willoughby Run. This was probably about twelve o'clock and about the time the other brigades of the I corps were reaching the field.

When the cavalry was relieved by the infantry it was withdrawn to the rear and then posted on the left flank of the line. Calef's battery was withdrawn to Seminary Ridge and enfiladed the railroad cut, but upon Davis' retreat, returned for a short time to McPherson Ridge. Although Heth had the brigades of Pettigrew and Brockenbrough in reserve he did not employ them to support Archer and Davis because, as he says in his report, his operation was simply an armed reconnaissance.

INTERVAL BETWEEN MORNING AND AFTERNOON ENGAGEMENTS.

After the defeat of Archer and Davis, the Union line was rectified by General Doubleday. Meredith's brigade was deployed along Reynolds Ridge in rear of the McPherson woods with a skirmish line in ~~it~~ them. Cutler was deployed along the same ridge north of the railroad, the gap between the two brigades was filled by Stone's brigade, which had come up about the end of the engagement and advanced to McPherson Ridge. Meredith's line was prolonged to the left by Rowley's brigade, which also arrived towards the end of the engagement, having marched from Marsh Creek bridge on the Emmitsburg road along Willoughby Run to the Fairfield road. From that point it moved to the general line and was deployed along Reynolds Ridge some distance north of the Fairfield road. Gamble's cavalry was assembled in rear of Seminary Ridge south of the Fairfield road. The corps artillery was posted in the best positions possible along Seminary Ridge. Baxter's and Paul's brigades of the second division were on their arrival posted in reserve at the Seminary itself. Shortly thereafter, Baxter's brigade was sent to the intersection of the Mummasburg road with Oak Ridge, to protect the right flank. As he did not connect with Cutler's brigade on his left, the division commander sent Paul's brigade to fill the gap. This brigade was posted in rear of a stone wall which followed the crest of the ridge north of Shead's grove, along what is now *Doubleday Avenue*.

General Howard, who had preceded his corps, arrived on the field during the morning fight and succeeded General Reynolds in command of the entire field. General Schurz took command of the XI corps. After consultation with General Doubleday, General Howard decided to send two divisions of his corps to prolong the right wing of the I corps along Oak Ridge and prevent its being turned by the Mummasburg road. Before his corps reached the ridge, however, Rodes' division appeared in force and the four brigades of Barlow's and Schurz's divisions of the XI corps, took up a line on the low ridge extending from Barlow's Knoll to Seminary Ridge.

General Howard recognized the importance of Cemetery Hill, and upon their arrival in the afternoon posted the two brigades of Steinwehr's division in reserve at that point where he also established his headquarters. Of the corps artillery, two batteries were posted north of the town near the Carlisle road; one was posted on the right of the line, four of its guns on Barlow's Knoll; one was held in reserve near the town; and the other was held in reserve on Cemetery Hill.

Devin's cavalry brigade, which was relieved by the XI corps, assembled east of Rock Creek near the Hanover road. In the afternoon it moved through the town to Gamble's position, just before the town was captured.

THE AFTERNOON ENGAGEMENT.

Before the battle was renewed in the afternoon, General Hill, with Pender's division, had reached the field from the west, and General Ewell with Rodes' division, from the north. At the opening of the battle Early's division had not arrived.

The infantry attack, which began probably about 2.30 P.M., was preceded by artillery fire. Two batteries of Rodes' divisional artillery opened on Cutler's and Stone's brigades from the west side of Oak Hill and compelled Cutler to fall back to Shead's grove and Stone to change front with two of his three regiments, from west to north; the other was concealed by the McPherson barn. The other two batteries of Rodes' division were sent to the plain east of Oak ~~Ridge~~ Hill. Pender's batteries were placed on the ridge west of Willoughby Run between the Cashtown and Fairfield roads.

In the afternoon of July 1 the brigades of Rowley and Meredith were attacked by the divisions of Heth and Pender; the brigades of Cutler, Paul and Baxter were attacked by those of O'Neal, Iverson, Daniel and Ramseur; Stone's brigade, being at the angle, was engaged both by Daniel and Brockenbrough; the first and third divisions of the XI corps were attacked first by Dole's brigade, and later by that brigade reinforced by Early's division.

General Hill, in making his attack, placed Heth's division in the first line, and Pender's in the second. Heth formed his brigades also in two lines, placing the brigades of Archer, Pettigrew and Brockenbrough in the first line and Davis in reserve. Archer, being threatened in flank by Gamble, did not attack. As a result only two of Heth's brigades were in contact with the enemy, while General Doubleday had all three of his brigades in the front line. Even Brockenbrough did not attack simultaneously with Pettigrew; the latter therefore struggled alone for some time with Meredith and Rowley. Doubleday was therefore able to check the advance of Hill's corps until about 3.45 P.M., when Pender's division was ordered to the front and the brigades of Scales, Perrin and Lane were formed for the attack. With this new line Hill was able to carry Seminary Ridge, but not before the escape of most of the troops, whose line of retreat would have been compromised by its earlier loss. The brigades of Meredith, Stone and Rowley retreated to Cemetery Hill, where they were posted by General Hancock. The retreat was protected by Buford, who first deployed Gamble's brigade on Seminary Ridge south of the Fairfield road to cover the left flank, and later reunited his command and threatened to take in flank any force crossing the valley in his front.

General Rodes, in making his attack, placed O'Neal's brigade across Oak Ridge and extended his line to the right by Iverson's brigade. Daniel's brigade was in echelon in rear of Iverson, and Ramseur in reserve in rear of Daniel. In attacking the right wing of the I corps it was necessary for Rodes' division to make a complete wheel to the left. This brought Iverson in the angle between Shead's grove and Oak Ridge and exposed him to the front fire of Paul's brigade from the stone wall and to the enfilade fire of Cutler. Rodes' attack was therefore wholly unsuccessful. O'Neal was repulsed by Baxter, who had the cover of a sunken road; Iverson was badly cut up by Cutler and Paul; and Daniel was diverted from his point of attack by Stone's brigade, which seized the opportunity, moved up to the railroad cut and took him in flank. He therefore became engaged with Stone over the possession of the cut. Doubleday's right wing easily held its line, until the retreat of the XI corps on its right compelled it to fall back. The troops then retreated along the foot of the ridge to the town and with the remainder of the I corps went to Cemetery Hill.

Doles' brigade of Rodes' division was for some time engaged alone with the XI corps and was obliged to call to its assistance one of the batteries from the west slope of Oak Hill. About 3 P.M., however, Early's division appeared on the Heidlersburg road on its left. Early at once established twelve guns to fire on Wilkeson's battery on Barlow's Knoll and enfilade Barlow's line. At the same time he deployed his brigades for attack; Hays' brigade across the road, Gordon on his right, Hoke on his left, and Smith in reserve. In this formation he advanced; Gordon assaulted and captured Barlow's Knoll, while Hays and Hoke advanced on his left to cut off the retreat of the XI corps. At this stage General Howard sent Coster's brigade from his reserve along the Heidlersburg road to the edge of the town to check the advance of Hays and Hoke and give time for his line to fall back to Cemetery Hill. The Union forces were now all falling back, and while there was much disorder in the town and many of the I and XI corps were captured, still the retreat was sufficiently orderly to permit a new line to be established on Cemetery Hill, overlooking the town, with Colonel Smith's Union brigade of the XI corps as its center.

The retreat of the Union troops was followed by the advance to the town of all of Ewell's brigades, save one; General W. Smith's Confederate brigade remained on the York road, probably because Kilpatrick was reported moving in along that road. Although in some confusion, both Early and Rodes awaited orders to attack Cemetery Hill. About this time General Lee reached Seminary Ridge and sent word to General Ewell to take Cemetery Hill, if he felt able to do so. General Ewell saw that the position had already been prepared for the defense and was held by a strong line of infantry and artillery. As he could not use his artillery against it, and as his troops had been either marching or fighting all day, he decided to delay the attack until the arrival of Johnson's division, which was momentarily expected; in the meantime he would reconnoiter the position. As Johnson did not come up before dark, the attack was not renewed.

Not receiving orders to move on Cemetery Hill, General Hill remained with Heth's and Pender's divisions on Seminary Ridge.

That night Heth and Pender bivouacked on the west slope of Seminary Ridge; Rodes along the Fairfield road between the Seminary Ridge and the town; Hays and Hoke (Avery) in the fields east of the town; and Gordon, W. Smith and Jenkins' cavalry east of Rock Creek.

In the battle of July 1 the I corps bore the brunt of the attack and was assailed by at least twice its numbers. Rodes gives the strength of his infantry at Carlisle as 8,065 present for duty. Excluding Doles' brigade he had 6,271 infantry and a battalion of artillery. (W. R. No. 44, p. 564.) The divisions of Heth and Pender of Hill's corps, with their artillery battalions, must have brought the number available for attacking the I corps to at least 20,000. The strength of the I corps is given in the returns of June 30 as 10,355 present for duty.

The losses were correspondingly heavy. The casualties of this corps in the battle of Gettysburg, most of which occurred on July 1, were:

	Killed.	Wounded.	Captured and Missing.	
1st division.....	299	1,229	627	
2nd division.....	91	616	983	
3rd division.....	265	1,297	541	
Total	655	3,142	2,151	
				5,948

33.6 per cent. in killed and wounded.

The XI corps had present for duty on June 30, 10,576, of which only two-thirds, or about 7,000 men, were deployed along the line in advance of the town. These were attacked by Doles' brigade and Early's division, probably about 9,000 in all. Their position was rendered untenable by the direction of Early's advance. The losses of the two advanced divisions in the battle of Gettysburg, most of which occurred on July 1, were also heavy.

	Killed.	Wounded.	Captured and Missing.
1st division.....	122	677	507
3rd division.....	105	507	332

CHAPTER VI.

BATTLE OF JULY 2.

MOVEMENT OF TROOPS NOT ENGAGED IN BATTLE OF JULY 1.

Confederate Army.—Early in the morning of July 1, Anderson's division of Hill's corps marched from the vicinity of Fayetteville to Cashtown in pursuance of orders received on the previous day. At the latter place, General Anderson heard the artillery of Heth's morning attack and an hour later received orders from General Hill to move on to Gettysburg. The division reached the field in the afternoon shortly after Pender's division moved to attack and remained in reserve until the close of the day.

On the morning of July 1, Johnson's division of Ewell's corps, which was still west of Fayetteville, started for Cashtown with Ewell's trains. In the vicinity of Fayetteville, it encountered McLaws' division, which was about to move to the same place. General Lee, who was present, gave Johnson the right of way and he followed Anderson through the gap. The movement of his division through that of McLaws caused some delay and it was evening before the head of Johnson's column reached Gettysburg. During the night his troops were moved across Rock Creek and bivouacked in the fields north of the Hanover Road.

The divisions of McLaws and Hood of Longstreet's corps followed Johnson through the gap. McLaws' advance was so delayed by Ewell's wagons that he bivouacked that night near the Marsh Creek bridge on the Cashtown road, about two miles from Willoughby Run. Here Hood joined him at midnight. Hood had left one brigade, Law's, on picket at New Guilford, which did not reach him until noon of July 2. Pickett's division was unable to leave the vicinity of Chambersburg until the morning of July 2 and reached Marsh Creek bridge that night.

On July 1 Stuart with the brigades of Fitz Lee and Chambliss marched from Dover, in the vicinity of York, to Carlisle; Hampton stopped at Dillsburg. During the night Stuart received orders to move on Gettysburg. Hampton's brigade, which was in advance, encountered Kilpatrick on the evening of the 2nd near Hunterstown. On the morning of July 3, Stuart reached Gettysburg with his three brigades and about noon moved towards the extreme left of the Confederate line.

Robertson's two brigades reached Chambersburg on the evening of July 2 and bivouacked on the road to Cashtown; on the following day they marched through Cashtown to Fairfield.

Imboden reached Chambersburg on the evening of July 1 and relieved Pickett; the next day he started for Gettysburg with the trains.

General Lee first heard the guns of Gettysburg as he rode towards Cashtown at the head of Longstreet's corps about noon of July 1. Hastening to the field he reached Hill's corps about 2:30 P.M.

UNION ARMY.—General Meade was at Taneytown on July 1, and received his first information of the battle about

1 P.M., when he learned of the death of General Reynolds. At this time the II corps was at Taneytown and he was explaining his proposed position on Pipe Creek to its commander, General Hancock. Since General Hancock was fully acquainted with his plans, he sent him at once to Gettysburg to take command of the I, III and XI corps and to examine the site. If under the existing circumstances General Hancock found the field a better one (than Pipe Creek), he was to notify General Meade, who would order up the remaining troops. (W. R. No. 45, p. 461.)

General Hancock proceeded to Gettysburg and arrived about the time the Union forces were falling back. He at once established a new line along the crest of Cemetery Hill, extending to the east and south. About 5:30 P.M. he sent a written report to General Meade, in which he reported that he had taken up a position in the cemetery which could not well be taken. It was a position, however, which could be readily turned. He reported General Slocum coming up and probably General Sickles. He stated that the ground appeared to him not unfavorable for a battle with good troops. (W. R. No. 45, p. 366.)

Upon the receipt of this communication, General Meade decided to concentrate at Gettysburg, towards which point many of his troops were already marching.

The II corps started towards Gettysburg early in the afternoon on the Taneytown road, to protect the flank of the I corps should it retreat to Emmitsburg. This corps reachd a point three miles south of Gettysburg after dark and there bivouacked for the night.

The III corps was ordered to the front on the afternoon

of July 1, by General Howard. This order was modified by General Meade, who directed that a division be left at Emmitsburg to guard that point. General Birney left one brigade at Emmitsburg and with the other two marched directly towards Gettysburg, reaching the left of the field after dark. General Humphreys also left one brigade at Emmitsburg and with the other two marched on a parallel road to the westward which entered the Emmitsburg road at the Peach Orchard. Being led astray by his guide, General Humphreys did not reach the field until 1 A.M., July 2. As soon as General Meade decided on the concentration at Gettysburg, he ordered the two brigades of the III corps, left at Emmitsburg, to join their divisions. They reached the field about 9 A.M. July 2.

The V corps, which had reached Hanover about 5 P.M., received orders that night to move towards Gettysburg. The first and second divisions made a night march and reached Bonneauville, or Bonaughton, at midnight; they were on the march again at 4 A.M., and reached the enemy's position, on the Hanover road, at 7 A.M. The third division joined the other divisions about noon.

In the afternoon of July 1, before ordering the concentration, General Meade decided that the VI corps was too far east and ordered it to move to the Taneytown road. Later he sent orders to the corps to move straight to Gettysburg. This corps made a forced night march and reached the field between 4 and 6 P.M., July 2, by the Baltimore pike.

The XII corps reached Two Taverns, on the Baltimore pike, five miles from Gettysburg, at 11 A.M. on July 1. In the afternoon General Slocum received notice from General

Howard that Hill and Ewell were both marching on Gettysburg and that the left wing of the army was engaged. At 3:30 P.M., General Slocum moved to Gettysburg and reached the field at 5 o'clock. The first division, on approaching Rock Creek, was moved to the Hanover road and threatened Ewell's left; it was later bivouacked on the Baltimore pike just east of Rock Creek. The second division crossed the creek and bivouacked near the Round Tops.

On the morning of July 2, General Buford, with the cavalry brigades of Gamble and Devin, was ordered to Taneytown to collect the trains left there and take them to Westminster; he left the field about 9 A.M. General D. McM. Gregg, with the brigades of Irvin Gregg and McIntosh, was ordered to Gettysburg and reached the field about noon; he was posted on the extreme right of the line on the Hanover road.

Kilpatrick's division spent the 1st and 2nd of July in reconnoitering the country north of the York road and on the evening of the 2nd encountered Hampton's brigade at Hunterstown. That night his division bivouacked at Two Taverns.

General Meade remained at Taneytown until he had dispatched the final orders for concentration to the III, V and VI corps. He then rode to Gettysburg and reached Cemetery Hill shortly after midnight July 1.

RESUMÉ.

On the morning of July 2, General Lee had, in the immediate vicinity of Gettysburg, all of Ewell's and Hill's corps, and two divisions of Longstreet's corps less one brigade, or thirty-three infantry brigades in all. Of these sixteen had not been engaged in the battle of July 1. He had also one brigade of cavalry.

At the same time General Meade had in the vicinity of Gettysburg all of the I, II, XI and XII corps, four brigades of the III and six brigades of the V corps. In all he had thirty-nine infantry brigades, of which only twelve had been engaged in the battle of the 1st. In addition he had Buford's two cavalry brigades. Although numerically stronger on paper, it is probable that there was little difference in the numerical strength of the two forces, as the Union army had been making long marches in a friendly country since June 28th, and there were many stragglers. The Confederate army, being in a hostile country, had few stragglers.

At noon Law's brigade had joined Longstreet's corps, the two brigades left at Emmitsburg had joined the III corps, and its third division had joined the V corps; Lee therefore had thirty-four brigades and Meade forty-three. During this morning also Gregg's two cavalry brigades and the reserve artillery of the Union army reached the field, but Buford's two brigades marched to Taneytown.

About 4 P.M. the VI corps began to arrive, and at dark all the infantry of both armies was on or near the field.

BATTLEFIELD OF JULY 2 AND 3.

The field occupied by the opposing corps on the 2nd and 3rd of July lies south of the *Fairfield* and the *Hanover* roads.

Its western limit is *Willoughby Run;* its eastern limit

is *Rock Creek*, from a point three miles south of Gettysburg northward to the bridge on the Baltimore pike; thence the eastern limit is the wooded ridge which passes through *Wolf Hill* and crosses the Hanover road about a mile east of Rock Creek. The field is traversed by three of the highways over which the Union army reached the field; the *Emmitsburg*, the *Taneytown* and the *Baltimore roads*. These are connected by various cross roads, the principal of which is the *Wheatfield road*, running from the *Peach Orchard* on the Emmitsburg road, eastward to the *Taneytown* road, where it connects with a road running to the bridge over Rock Creek on the Baltimore pike. From the Peach Orchard the road runs westward to Willoughby Run, whence a branch road runs to Black Horse Tavern on the Fairfield road.

The field is traversed by various ridges which separate the watersheds of Rock Creek, Willoughby Run and their various tributaries.

The principal of these ridges are *East Cemetery and Culp's Ridge, Cemetery and Round Top Ridge, Emmitsburg Road Ridge, Seminary Ridge and Snyder's Ridge.*

East Cemetery and Culp's Ridge, now followed by *Slocum Avenue*, runs eastward from *Cemetery Hill* to *Culp's Hill* and thence parallel to the Baltimore pike to *Spangler's Spring.* Each of the hills commands the field to the north by about one hundred feet. The lowest point of the connecting ridge is about fifty feet lower than the summits of the hills. From Culp's Hill the crest of the ridge slopes gradually downwards to Spangler's Spring, where the ridge terminates. Culp's Hill and the ridge to the south was covered by a thick growth of timber and was strewn with boulders.

In rear of Culp's Hill is a valley followed by the Baltimore pike. This valley and the ridge south of Culp's Hill is commanded by *Power's Hill,* in the angle between the Baltimore pike and the School House road, and by the high ground between the Baltimore and Taneytown roads.

Cemetery and *Round Top Ridge*, now followed by *Hancock, Sedgwick* and *Sykes Avenues*, runs southward from Cemetery Hill to Big Round Top. From Cemetery Hill, as far southward as the intersection of Plum Run and the Emmitsburg road, near *Codori's* house, this ridge is very clearly defined and commands all the ground to Seminary Ridge on its west. From this point southward, as far as the Round Tops, it is broken, wooded and the field of view is limited by the ridge followed by the Emmitsburg road. *Big Round Top* is the most conspicuous elevation in the vicinity of Gettysburg. It is a conical peak, covered by a thick growth of timber and innumerable boulders, which rises one hundred and fifty feet above Cemetery Hill. *Little Round Top* is a rocky ridge about fifty feet higher than Cemetery Hill, and is separated from Big Round Top by a wooded depression about forty feet lower than Little Round Top. From Little Round Top the view is limited only by Seminary and Snyder's ridges. At the time of the battle the slope in front of Little Round Top was cleared of timber.

Snyder's Ridge is the prolongation of the ridge between the Cashtown and Fairfield roads just west of Willoughby Run. It crosses the Emmitsburg road about three-quarters of a mile south of the Peach Orchard.

Seminary Ridge, south of the Fairfield road, is a prolongation of the same ridge north of that road. It runs nearly parallel to the Emmitsburg road until it intersects Snyder's Ridge.

Seminary and Snyder's Ridges are now followed by *Confederate Avenue.*

The *Emmitsburg Road Ridge* begins just west of that road, near the crossing of Plum Run and terminates near the intersection of the Emmitsburg-Wheatfield roads. It is now followed by *Sickles Avenue.*

Devil's Den is a hill on the opposite side of Plum Run, in front of Little Round Top and commanded by it. *The Loop* is a rocky salient of the ridge between Devil's Den and the Peach Orchard and midway between them. *The Wheatfield* is an open field almost surrounded by woods which lies behind the line connecting Devil's Den and The Loop.

Brinkerhoff Ridge, east of Rock Creek, is a ridge extending from Rock Creek across the Hanover road; Wolf Hill, on this ridge, is about twenty-five feet higher than Culp's Hill. In the angle between Rock Creek and the Hanover Road and south of the road is *Benner's Hill,* whose elevation is fifty feet less than Culp's Hill.

At the time of the battle, Wolf Hill, Culp's Hill, Big Round Top, and nearly all the field in front of Big Round Top, Devil's Den and The Loop, were covered by trees and undergrowth.

Plum Run is a branch of Rock Creek, whose valley separates Devil's Den from the Round Tops; a branch valley also separates Devil's Den and The Loop from the Emmitsburg road. *Pitzer's Run* is a small branch of Willoughby Run, which is nearly parallel to Seminary Ridge, and has two small dry branches which intersect that ridge.

UNION LINE ON THE MORNING OF JULY 2.

General Meade arrived on the field shortly after midnight of July 1 and at once proceeded to study the position and decide on the proper disposition of the troops.

The XI corps was left in its position of the previous evening. Its artillery was on the crest of Cemetery Hill and its infantry on the slope in advance; one division west of the Taneytown road, one between the Taneytown and Baltimore roads, and one to the right of the Baltimore road.

The I corps, now commanded by General Newton, had one division, Wadsworth's, on the right flank of the XI extending the line towards Culp's Hill; the other divisions formed a reserve, in rear of the XI corps.

The second division of the XII corps, being relieved by the III corps, was marched from the vicinity of the Round Tops to Culp's Hill and there put in position on the right of the I corps. Its line extended from the summit of the hill nearly as far as Spangler's Spring; two brigades being in the first line and one in the second. The first division of the XII corps, being relieved on the east side of Rock Creek by the V corps, was posted on the right of the second division, extending the line along the creek. One brigade was to the left of Spangler's Spring and the others to the right. Five guns were placed on Culp's Hill in the afternoon of that day.

The first and second divisions of the V corps arrived on the Hanover Road, east of Brinkerhoff Ridge, at 7 A.M.; they remained east of Rock Creek until Gregg's cavalry division arrived in that vicinity about noon; the two divisions, now joined by the third, then took post as a reserve on the west bank of Rock Creek, where the roads fork to Cemetery Hill and Little Round Top.

The II corps marched to the field early in the morning and was posted on the left of the XI corps, extending the line towards Little Round Top; the third division on the right, the second in the center, and the first on the left. Each division had one brigade in the second line. The corps artillery was placed in the intervals between the brigades of the first line.

In the morning, the III corps was bivouacked in mass along the Taneytown road, north of Little Round Top. Being ordered to relieve Geary's division of the XII corps and extend the line of the II corps southward, the first division was extended to Little Round Top. In this position the corps was joined, between 9 and 10 A.M., by the brigades which had been left at Emmitsburg. About noon the second division was moved to a line in front of Plum Run and deployed in three lines. The front of the corps now extended from Little Round Top to a point in front of the II corps. The picket line of the corps was along the Emmitsburg road.

The VI corps did not reach the field during the morning.

CONFEDERATE POSITION.

Because of the absence of his cavalry, General Lee was wholly in ignorance as to the positions of all the corps of the Union army, save the I and the XI corps. As the Union line from Cemetery Hill eastward appeared to be a strong one, on the evening of July 1 General Lee ordered General

Ewell to move his corps to the right of Hill, if it could not be used to advantage where it was. Two of Ewell's aids had, however, been on Culp's Hill in the early evening and reported it unoccupied; as this hill was an important tactical point, Ewell was allowed to remain to secure it. At midnight, therefore, Ewell directed Johnson to occupy this hill if he had not already done so; but the latter, on sending out a reconnoitering party, found it occupied by a regiment of the I corps. While the reconnoitering party was out, Johnson captured a message from General Sykes to General Meade, saying that he would reach the field at daylight by the Hanover road. As this would bring the V corps on his flank and rear, Johnson decided to await further orders before attacking Culp's Hill.

General Lee had decided to make his main attack on July 2 with his right wing, as the ground here seemed favorable for offense. The main attack was to be made by Longstreet with the three divisions not employed on the preceding day; those of McLaws, Hood and Anderson. As it was too late to move Ewell's corps to their support after the latter found he could not take Culp's Hill without fighting, General Lee ordered him to make a demonstration at the time of Longstreet's attack, and convert it into a regular attack if possible. General Hill was to occupy the center and threaten attack, to prevent the reinforcement of the wings of the Union army.

In the morning, the position of the Confederate army was as follows: East of Rock Creek were Johnson's division of four brigades north of the Hanover road, and Gordon's and Smith's brigades of Early's division near the York road.

Between Rock Creek and the town were the brigades of Hoke and Hays. In the town, and extending along the Fairfield road to Seminary Ridge, were the brigades of Doles, Iverson and Ramseur; to their right and rear, along the railroad, were those of Daniel and O'Neal. Pender's division was along Seminary Ridge, its left resting on the Fairfield road. Heth's division was near Pender's; Anderson's division was in rear of Willoughby Run; McLaws' and Hood's divisions were near Marsh Creek on the Cashtown road.

It was General Lee's intention to make the attack in the morning as early as possible, but there were unavoidable delays due to the necessary examination of the position and the movement of the troops into place.

In the morning Anderson's division was moved from Willoughby Run and deployed in a single line along Seminary Ridge south of Pender's division, thus relieving Heth's division, which was moved back to Willoughby Run. The brigade of Wilcox, which formed his right, was beyond Spangler's grove. In reaching this position, this brigade, about 11.45 A.M., encountered details from two regiments of sharp-shooters and the 3rd Maine infantry, sent by General Sickles on a reconnaissance. At the time, these troops were in the woods on Snyder's Ridge south of Pitzer's Run. The Union troops were compelled to fall back to the Emmitsburg road.

In the morning, Longstreet's divisions were moved from Marsh Creek to a position near Black Horse Tavern, but north of the Fairfield road. About noon, being joined by Law's brigade, they started to march by a road leading from Black Horse Tavern direct to Willoughby Run. After moving for some distance along this road it was apparent that the column would come in full view of Big Round Top. As it was desired to have the attack in the nature of a surprise, the column was countermarched to a point midway between the Cashtown and Fairfield roads, moved into the valley of Willoughby Run from that point, and then down the valley to the right of Anderson's division. Here the divisions were deployed in line behind the belt of woods which crowned Snyder's Ridge. In the front line, from right to left, were the brigades of Law, Robertson, Kershaw and Barksdale; in the second line those of Benning, Anderson, Semmes and Wofford. The center of McLaws' division was the road leading to the Peach Orchard; that of Hood, the Emmitsburg road. Alexander's battalion of corps artillery was in the center of McLaws' division, and Cabell's on his right. One-half of Anderson's divisional artillery was between Wilcox and McLaws. Two batteries of Henry's artillery were on the extreme right of the line. The development of the troops for battle was not completed until 4 P. M.

CHANGE OF POSITION OF THE III CORPS.

On the morning of July 2, General Meade's headquarters were near the Cemetery and he was occupied almost wholly in establishing the lines of the I, II, XI and XII corps. He proposed an attack on Ewell's corps by the XII corps, reinforced by the V corps, but upon reconnoitering the position, General Slocum, commander of the XII corps, and General Warren, the chief engineer of the army, both reported against it. At this time all of the Confederates visible in the field were in front of the positions occupied by the I, II, XI and XII corps. All of Buford's cavalry had, by a misunderstanding, been sent away from the left of the line by General Pleasanton, a fact that General Meade was not aware of until informed by General Sickles about noon; orders were at once sent to have it replaced. This order could not be complied with, however, until the following day. For information as to the movements of the Confederates on the left, General Meade had to rely on the signal station on Little Round Top. The first report sent by the signal station of operations on the left flank was made at 11:45 A.M., when Wilcox's skirmishers were sighted. At 11.55 A. M. the signal station reported the III corps skirmishers falling back. (W. R. No. 45, p. 487.)

The movement of Longstreet's corps towards the Union left flank was apparently not discovered by anyone, until his skirmishers appeared in the woods along Snyder's Ridge at 4:00 o'clock. At 1:30 P.M. the signal station reported a large force, 10,000 strong, moving from the extreme left to the right, and at 2:10 P.M. added that the troops were moving towards Herr's tavern on the Cashtown road. (W. R. No. 45, p. 488.) This column was evidently Longstreet's making its countermarch.

On the morning of July 2, the III corps was bivouacked in mass east of Plum Run near what is now the junction of *Hancock* and *United States* avenues; Humphreys' division was on the right, Birney's on the left. On the right of Humphreys was Caldwell's division of the II corps; on the left of Birney was Geary's division of the XII corps. Some of Geary's troops were on Little Round Top, but the division was in bivouac rather than in position.

The instructions sent to General Sickles by General Meade were to form line of battle on the ground occupied by Geary's division connecting on the right with the II corps and extending the left to the Round Tops if practicable. In the morning therefore when Geary moved away and went to Culp's Hill, Birney deployed his division to cover the front between Humphreys' division and Little Round Top. The two brigades left at Emmitsburg joined the corps between 9 and 10 A.M., and shortly thereafter a staff officer came from General Meade's headquarters to see if the III corps was in position.

The ground occupied by the III corps was rough and wooded; to its front were the woods extending from Devil's Den along the farther side of the Wheatfield and thence in a fringe eastward along Plum Run. It was commanded by the Emmitsburg road ridge, which limited the view in this direction. In front of the Emmitsburg road ridge, where the III corps had its skirmish line, was open country in all directions as far as Seminary and Snyder's ridges.

Being impressed by the defects of his position and the advantages of the Emmitsburg road ridge, General Sickles rode to General Meade's headquarters to request General Meade either to examine the ground in front of the III corps himself or to send his chief engineer, General Warren.

Unfortunately General Meade did not anticipate an attack on his left and did not feel that either he or Warren could go. However, at that moment General Hunt, chief of artillery, came in, and he was requested by General Meade to accompany Sickles. It was 11 A.M. when the two officers rode towards the Peach Orchard to examine Sickles' proposed line. General Hunt was impressed by the favorable position it would give the enemy if abandoned to him, but also by the difficulties attendant on its occupation by the III corps. He saw that the right flank of the III corps would be in the air and that its front would be too long for its effective strength. He also noted the woods on Snyder's Ridge in front of the Peach Orchard and suggested that a reconnoitering party be sent into them. It was this party that met Wilcox about noon.

General Hunt would not take the responsibility of authorizing General Sickles to move to his proposed line, but said he would report the matter to General Meade. While awaiting the orders he hoped and expected to receive, General Sickles advanced Humphreys' division beyond Plum Run and deployed it in three lines, one brigade in each.

As Anderson was now in force in the woods in his front, it was apparent to General Sickles that he must either occupy the Emmitsburg road ridge at once or abandon all hope of occupying it, as the Confederates would soon drive back his skirmish line. Believing he would be sustained in his action he therefore advanced his whole line.

Ward's brigade of Birney's division was advanced to Devil's Den and the woods to its right, and a battery was placed on the Den. De Trobriand's brigade was advanced to the Loop and the woods in that vicinity. Graham's brigade was advanced to Peach Orchard cross roads. As Birney's line was long, he was given Burling's brigade of Humphreys' division as a reserve. The movement was completed by advancing Humphreys' division to the Emmitsburg road. Along the Emmitsburg road were placed three batteries of artillery, and a fourth was placed along the Wheatfield road east of the cross roads. The general line of the III corps is now marked by Sickles Avenue. The troops had just reached their new positions when Longstreet's artillery appeared in their front.

General Meade was not aware of this movement of the III corps, since General Hunt had reported to him that he had advised General Sickles not to move his troops unless authorized by General Meade. A meeting of the corps commanders was ordered for the afternoon of July 2 in anticipation of the arrival of General Sedgwick with the VI corps. General Meade had already ordered the V corps to move to the left of the line to make way for the VI corps. As General Sickles approached General Meade's headquarters, Longstreet's artillery opened fire. He at once returned to his corps and was followed by General Meade. The latter appreciated the impossibility of withdrawing the III corps under fire and therefore confined himself to reinforcing it. General Sykes was ordered to support Sickles with the V corps and Caldwell's division of the II corps. When at about 6 P.M., the whole advance line gave way, he ordered up the XII corps.

General Hunt was directed to assist Sickles with artillery and at once sent forward McGilvery's brigade, of which two batteries were posted in the Peach Orchard, and two on the Wheatfield road to its right and rear facing south.

THE CONFEDERATE ATTACK.

General Lee, whose headquarters were at the Seminary,

had not reconnoitered the field of attack since morning. He was under the impression that the main Union line lay along the Emmitsburg road. His plan was therefore to form Longstreet's divisions across this road beyond the Union left flank and then sweep forward along the road. Although the conditions were found by Longstreet, who had personal charge of the attack, to differ somewhat from those upon which the movement was based, he did not feel justified in changing the plans already made. After the preliminary artillery attack, the brigades were to move forward in echelon from right to left and then each face to the left and move parallel to the Emmitsburg road.

The artillery opened about 4 P.M. and immediately thereafter Law's brigade moved forward; the movement was carried along very slowly to the left, which did not move until about 6 P.M.

Law's brigade, followed by Robertson's, moved through the open fields in front of the extreme right of the Confederate line south of Devil's Den. At Devil's Den the left two regiments of Robertson's brigade became involved with the troops behind a stone wall in the woods at its foot, while the others moved on with Law's brigade to attack Little Round Top.

About 4 P.M. General Warren, who had accompanied General Meade to Sickles' position, ascended Little Round Top, which was then unoccupied by troops, and caught the reflection of the bayonets of Hood's division as it was preparing to move. Being impressed by the importance of the hill, he at once hurried off to meet the advancing V corps and secured Vincent's brigade for its protection. Later when the next division arrived, Weed's brigade and Hazlett's battery were sent to the same position. Vincent was hardly in position when Law's column attacked the hill in front and on the left flank; desperate fighting here continued until dark, but the Confederates were continually repulsed. After dark some regiments of Vincent's and Fisher's brigades seized and occupied Big Round Top. At Little Round Top both of the brigade commanders, Vincent and Weed, and the battery commander, Hazlett, were killed.

Robertson, with his two left regiments, encountered a vastly superior force at Devil's Den and several times sent back for assistance. At last the brigades of Anderson and Benning came to his relief. This force drove Ward's brigade from the hill and captured most of the guns.

Kershaw's brigade moved from its position in line directly on the Loop, occupied by De Trobriand's brigade. In moving forward, however, it received the fire of some Union troops in the Peach Orchard and from the batteries along the Wheatfield road. This diverted his left regiments to their left and he reached the knoll too weak to carry it and was driven back. Some time later Semmes' brigade came up on his right and Wofford's on his left and together they carried the Loop. The Confederates now had a continuous line extending from Devil's Den diagonally to the Wheatfield road. This line now wheeled gradually to the right and ended the day by an unsuccessful assault on Little Round Top and the ridge to its north. (Battles and Leaders, Vol. III page 331)

Longstreet's artillery having silenced that of the III corps, about 6 P.M. Barksdale's brigade moved over the field to the north of the cross roads and carried all before him until he reached Plum Run. This advance was participated in by the brigades of Wilcox, Perry and Wright on his left. With the exception of Wright's brigade, all of these brigades were stopped near Plum Run; Wright's brigade carried the main Union line in its front, but having no support was compelled to retire.

THE DEFENSE.

The III corps fought desperately before falling back; General Sickles was severely wounded early in the fight and General Meade ordered General Hancock to assume command of the II and III corps. The Union reserves were sent in wherever weakness was developed in the Union line. The first and second brigades of Barnes' division of the V corps were sent to reinforce de Trobriand's brigade at The Loop; the first and second brigades of Ayres' division of regulars of the V corps were sent to reinforce Ward's brigade and cover the gorge between Devil's Den and Little Round Top; Caldwell's division of the II corps was pushed into the Wheatfield between the two. All of these troops were, however, compelled to fall back when the brigades of Barksdale, Wilcox, Perry and Wright turned their right.

Willard's brigade, which was in the second line of the II corps on the afternoon of July 2, was moved to the left of the line and there made a countercharge on Barksdale's command to recapture some guns in his rear. He encountered Barksdale's brigade in the underbrush along Plum Run; Barksdale was killed and his brigade was driven back towards the Emmitsburg road. In this movement Willard was also killed.

From Harrow's brigade, which was also in the second line of the II corps, two regiments were sent about two hundred yards in front of the general line to protect the right flank of the III corps; the other regiments were moved to the left to protect the batteries left uncovered by Caldwell's withdrawal from the main line. It was one of these regiments, the 1st Minnesota, which was selected by General Hancock to make a countercharge on Wilcox's brigade as it was emerging from the underbrush along Plum Run near the close of the battle; the percentage of loss suffered by this regiment in this charge and upon the following day was greater than that of any other Union regiment engaged in the battle.

Farther to the left, towards the close of the day, countercharges were made by Crawford's division of the V corps and later about dark by Lockwood's brigade of the XII corps.

When the fighting closed for the day the brigades of Hood's division held an advanced line along the slope of Big Round Top midway between its summit and Plum Run. McLaws' division held the woods from Devil's Den to the Wheatfield road. Longstreet's artillery had advanced to the Emmitsburg road. The Union line was that now marked by *Hancock, Sedgwick* and *Sykes* avenues and was that occupied on July 3.

EWELL'S ATTACK.

Ewell's corps was to attack simultaneously with Longstreet's. It was difficult for this corps to act as a unit, as it was separated into three parts by the town and by Rock Creek. As a result the three divisions acted independently.

When he heard Longstreet's guns, Johnson established his divisional artillery upon Benner's Hill and opened fire upon Culp's Hill. These batteries, being exposed to the concentrated fire of the Union guns on Culp's and Cemetery Hills, at ranges of one thousand and one thousand four hundred yards, suffered severely and were eventually withdrawn. The infantry was then formed for the attack parallel to the ~~York~~ Hanover Road and was about moving forward when Gregg's cavalry brigades attacked Walker's pickets on the Hanover road; Walker's brigade was therefore detached to meet him. The brigades of Steuart, Nichols (Williams) and Jones, however, continued on.

When the III corps was about to fall back, General Slocum was ordered to send the XII corps to its relief. Not wishing to abandon Culp's Hill while threatened by Johnson, General Slocum retained Greene's brigade which was posted on its summit. All the other brigades of the XII corps abandoned their trenches to go to the point of attack. Only Lockwood's brigade, however, reached that point in time to take part in the action.

Some time after the XII corps was withdrawn, the brigades of Steuart, Nichols (Williams) and Jones, of Johnson's division, crossed Rock Creek and took possession of the vacated trenches. Only when they attempted to ascend Culp's Hill did they meet with any opposition. Here General Greene's brigade was protected by a breastwork of stones and logs and was able to hold this important position. Aid was sent him by the troops on his left. It was dark when the Confederates crossed Rock Creek and it was 10 P.M. when fighting ceased at this part of the line. Although Johnson's men were close to the ammunition train of the Union army near the Baltimore road they were unaware of this fact.

When the different brigades of the XII corps returned to their original positions during the night, they found the Confederates occupying the line from Culp's Hill to Spangler's Spring; they therefore bivouacked in the vicinity and waited for daylight to drive the Confederates back across the creek.

To the east of the town, Early had the brigades of Hays and Hoke in the first line and that of Gordon, which had returned to the town, in the second; Smith's brigade was still out on the York Road. Early was ordered to attack Cemetery Hill when Johnson attacked Culp's Hill. He moved forward about dusk and drove the XI corps infantry from the stone wall on the slope of the hill and Hays penetrated the batteries on the crest. The attack had by this time become so disorganized that the Confederate troops were easily driven out by troops sent from the left. In this attack Col. Avery, who commanded Hoke's brigade in this campaign, was killed.

When Rodes saw the commotion in the Union line caused by the withdrawal of troops and batteries to go to the left, he decided to attack and so notified Early. However, before Rodes could reach the enemy's line, it was necessary for him to withdraw his troops from the town, deploy on a new line parallel to the Emmitsburg road, and then advance over a long distance. His line had not reached the Emmitsburg road when he heard of Early's defeat. As it was getting dark he decided thefore to defer his attack and await orders. To be ready for an attack in the morning, he bivouacked his troops along a sunken road parallel to the Emmitsburg road and 600 yards from it.

COUNCIL OF WAR.

Being impressed probably by the insecurity of his flanks as well as by his losses, General Meade called for a council of his corps commanders on the night of July 2-3. At this council the following propositions were voted on:

1st—Shall the army remain in its present position or retire to another nearer its base?

2nd—If it remains, shall it attack the enemy or await the attack of the enemy

3rd—If it awaits attack, how long should it await such attack?

After some discussion the corps commanders unanimously voted to remain and await an attack for at least one day.

It was therefore decided to strengthen the position for defense and on the following morning await attack by General Lee.

CHAPTER VIII.

BATTLE OF JULY 3.

POSITIONS.

UNION LINE.—On the morning of July 3, the Union line extended from a point east of Rock Creek on the right, to a point beyond Big Round Top on the left.

On the extreme right, across Rock Creek, north of the Baltimore road, was Neill's brigade of the VI corps, along what is now *Neill Avenue.* Encircling the Confederate trenches along Culp's Hill ridge was the XII corps with Shaler's brigade of the VI corps. The corps artillery occupied Power's Hill and other commanding positions along the line. The positions of the XI corps, the first and second divisions of the I corps, and the second and third divisions of the II corps were as on the preceding day. To the left of the second division of the II corps was the third division of the I corps. The reserve artillery was next in line, its center opposite the School House Road; intrenched in its rear was the first division of the II corps. To its left were two brigades of the VI corps. The V corps extended the line to Big Round Top, and two brigades of the VI corps, along what is now *Wright Avenue,* protected the left flank. The III corps and the remaining brigades of the VI corps were in reserve. The artillery was posted along the line, wherever a good position could be found. On the left flank, forming a line from Willoughby Run to Big Round Top, were Merritt's and Farnsworth's cavalry brigades, under Kilpatrick. On the right flank, extending from Wolf Hill or Brinkerhoff Ridge to White's ~~Creek~~ Run, along the Hanover road, were the cavalry brigades of Irvin Gregg, McIntosh and Custer under General D. McM. Gregg.

CONFEDERATE LINE.—On the morning of July 3, Jenkins' cavalry brigade was east of Brinkerhoff Ridge between the Hanover and York roads. This brigade relieved the infantry brigades of Smith and Walker, which were sent to Johnson. Stuart's cavalry was approaching the field from Jenkins' rear.

Johnson was holding the trenches along Culp's ridge, which he had captured; he had been reinforced by the brigades of Walker and Smith from his rear, and by those of Daniel and O'Neal from Rodes' second line. The brigades of Hoke and Hays occupied the town, while Gordon's brigade occupied the ground between the town and Rock Creek.

The brigades of Doles, Iverson and Ramseur, of Rodes' division, and those of Perrin and Thomas, of Pender's division, held the sunken road in the valley extending from the town parallel to the Emmitsburg road.

The divisions of McLaws and Hood held the line south of the Wheatfield road, extending over Devil's Den to the slopes of Big Round Top and thence westward to Willoughby Run.

Along Seminary Ridge between Thomas' brigade on the left and McLaws' division on the right, were being assembled the divisions of Heth and Pickett in the first line, supported by the brigades of Lane, Scales, Wright, Perry and Wilcox in the second line.

Influenced by his successes of the previous days General Lee decided to make another effort to pierce the Union line. The attack was to be made simultaneously by Johnson, who had been reinforced by the three brigades of Daniel, O'Neal and Smith, and by the brigades that were being assembled under Longstreet along Seminary Ridge.

Before his troops were ready to attack, however, Johnson was himself attacked by General Slocum.

ATTACK OF THE XII CORPS.

At daylight the batteries of the XII corps, which had taken up commanding positions enveloping the Confederate position on Culp's Ridge, opened fire. To this fire Johnson's command could make no reply, since he had no positions upon which he could establish batteries. The artillery fire was followed by an infantry attack, in which Shaler's brigade of the VI corps also participated. Though Johnson made desperate efforts to hold his position, he was compelled to retire and about 11 A.M. retreated to the position from which he had advanced the preceding day. The XII then reoccupied its trenches.

ATTACK ON CEMETERY RIDGE.

Early in the morning the Confederate infantry and artillery began moving in place for the proposed attack on Cemetery Ridge. The center of attack selected by General Lee was a salient in the line of the II corps, marked by an umbrella-shaped clump of trees between the brigades of Webb and Hall of Gibbon's division. A line joining this with the salient angle of Spangler's grove on Seminary Ridge separated the artillery of Longstreet and Hill, and the divisions of Heth and Pickett. Both Generals Lee and Longstreet viewed the final charge from positions near this line.

Extending northward from the Peach Orchard cross roads, along Emmitsburg road ridge, Colonel Alexander, Longstreet's chief of artillery, placed seventy-five guns. Extending southward from the Fairfield road along Seminary Ridge General Hill had fifty-eight guns and in addition five more on the left of Longstreet's artillery. To oppose these one hundred and thirty-eight guns and sweep the field over which the attack was to be made, the II corps had twenty-six guns, the artillery reserve thirty-nine guns, and a rifled battery of six guns on Little Round Top: in all seventy-one guns. General Hunt, however, had many batteries in reserve. The average distance between the opposing lines was about three-quarters of a mile.

Pickett moved his command to the field and placed them in and about Spangler's grove. Heth's division was moved to the field and placed along Seminary Ridge from Spangler's grove northward to the grove in rear of the right flank of Rodes' advanced line in the valley. The formation for the attack is thus described by General Longstreet:

"Orders were given to Major General Pickett to form his line under the best cover that he could get from the enemy's batteries and so that the assaulting column would arrive at the salient of the enemy's position. General Pickett's line to be the guide and to attack the line of the enemy's defenses.

General Pettigrew, in command of Heth's division, moving on the same line as General Pickett, was to assault the salient at the same moment. Pickett's division was arranged with two brigades in the front line, supported by his third brigade. Wilcox's brigade was ordered to move in rear of Pickett's right flank to protect it from any force the enemy might send against it.

"Heth's division, under the command of General Pettigrew, was arranged in two lines, and these were supported by part of General Pender's division, under General Trimble."

At 1 P.M. the signal guns for the artillery bombardment were fired and at once the Confederate artillery opened fire. The instructions of General Hunt to the Union artillery were to fire slowly and deliberately in reply and not to waste ammunition. These instructions were followed by the artillery reserve, but in the II corps, to save his infantry from loss and demoralization, General Hancock directed his chief of artillery to reply vigorously.

To Colonel Alexander, General Longstreet had assigned the responsibility of deciding the moment when the infantry was to advance. It was about 1:30 P.M. when Alexander decided to give the order, lest his ammunition run too low to support the attack. Five minutes later the II corps artillery ceased firing because it had exhausted all its long range ammunition, and the artillery reserve ceased, to keep its ammunition for the infantry assault which was expected. A few minutes later Alexander urged Pickett to hurry, and the latter riding to General Longstreet to have the order confirmed, moved to the attack. When the order was given,

Pickett had formed his division in front of the grove, in the ravines and along the fences north and south of the Spangler house. Each brigade was in line; Garnett on the left, Kemper on the right, and Armistead in rear. Moving toward the point of attack they formed three long lines which could be enfiladed from the position of the artillery reserve.

Along the front of attack and about a hundred yards in front of the crest of Cemetery Ridge is a low stone wall prolonged to the south by a rail fence. In rear of this wall and fence from right to left were the brigades of Webb, Hall and Harrow, each with one regiment behind the crest in rear. Along the stone wall were some of the guns of the II corps waiting to receive the charge with canister, while on the crest in rear were some fresh batteries of the artillery reserve which had replaced those of the II corps.

To the right of Webb but two hundred and fifty feet to his rear was another stone wall parallel to the front which protected Smyth's brigade; Willard's was in his rear. A battery in Ziegler's grove on Smyth's right swept his front.

All of the attacking troops moved towards Webb and Hall except the brigades of Davis and Brockenbrough, who attacked Smyth.

As the attacking troops converged on this small front, the brigades intermingled and it was a mass of mixed units led by General Armistead, which finally swept over the wall along Webb's front. Nearly all of these men were killed or captured by the Union reserves in rear, or by Hall's and Harrow's men on the flanks.

Seeing that Pickett's column would strike the line on his right, Stannard formed two of his regiments in front and perpendicular to the general line and took Pickett in flank. He captured three regimental flags and many prisoners. Davis and Brokenbrough were repulsed by Smyth before they reached his wall.

Pickett's three divisions according to the returns numbered about 5,000 men; of these 1,389 were killed or wounded, and 1,499 were captured. Among the killed were two of his brigade commanders, Garnett and Armistead, and among the severely wounded, was the third, Kemper. The six brigades of Heth and Pender having lost severely on July 1, it is impossible to give their losses or numbers accurately; they probably arrived on the field of Gettysburg with over 10,000 men. Capt. Young of the 26th N. C. reports that Heth's division reached the field of Gettysburg with about 8,000 effectives and on the evening of July 3, had but 1,500 or 1,600 men. His own regiment of Pettigrew's brigade was reduced from 800 to 216 on the afternoon of July 1, and to 80 men fit for duty on July 3. A major was left to command the brigade and the regiment was commanded by the only captain fit for duty. (W. R. No. 44, p. 645.)

General Trimble, who commanded Pender's brigades, the latter officer having been wounded, was captured in the assault.

The brigades of Wilcox, Lang (Perry) and Wright moved out after Pickett. Those of Wilcox and Lang (Perry) moved towards the position of the artillery reserve; they reached Plum Run, but being subjected to a severe artillery fire and being threatened by a flank attack of Stannard's brigade, fell back to their original positions. Wright's brigade advanced only a short distance when it was recalled by General Longstreet.

After the repulse, the lines were re-formed behind the artillery.

There was no counter-attack and the day was closed by an armed reconnaissance by some brigades of the V and VI corps which advanced along the Wheatfield road to the Emmitsburg road, driving back the Confederate troops in their front.

CAVALRY OPERATIONS ON JULY 3.

OPERATIONS ON THE UNION RIGHT FLANK.

Union Cavalry.—At noon on July 2, the brigades of McIntosh and Irvin Gregg of General D. McM. Gregg's cavalry division, reached a point on the Hanover road about three miles east of Rock Creek, where the *Low Dutch*, or *Salem Church* road connects the York, Hanover and Baltimore roads. At this time a regiment of the V corps was on outpost duty on Brinkerhoff Ridge. At 3 P.M. this regiment was relieved by the cavalry and joined its corps. The Confederate outpost on the Hanover road at Brinckerhoff's Ridge was a regiment of Walker's brigade of Johnson's division.

At 6 P.M. a reconnoitering squad of fifty men was sent forward by the cavalry to develop the Confederate strength. This led to an engagement between the Confederate infantry and some squadrons of dismounted cavalry on the ridge near the *Howard* house, west of Cress' Run. As Walker's regiment was repulsed, he reinforced it with his whole brigade and was therefore absent that night from the attack of Culp's Hill. At 10 P.M. the cavalry retired to the Baltimore road at White's Run, a mile east of Rock Creek.

The orders for July 3, were for Gregg's division to move back to the Hanover road to protect the right flank of the army, and Kilpatrick's division, which bivouacked that night at Two Taverns, to replace Buford on the left flank. By mistake, Custer's brigade of Kilpatrick's division moved up the Low Dutch road to the Hanover road.

Until noon, Custer remained in the vicinity of the cross-roads with outpost squadrons about a mile and a half north of it, on both the Low Dutch road and the intersecting road connecting the Low Dutch with the York road. Gregg's brigade was massed about a quarter of a mile south of the Hanover road and about the same distance west of Cress' Run. He had established a skirmish line extending from the Hanover road in front, westward to Wolf Hill where it connected with that of Neill's brigade of the VI corps. McIntosh's brigade was massed along the Low Dutch road about a half mile behind Custer. Two of his regiments, the 1st Pennsylvania and the 1st Massachusetts, were absent on detached duty. He had been reinforced by a company of Maryland cavalry and a platoon of two guns of the 3rd Pa. heavy artillery picked up in the march to the field.

About noon General D. McM. Gregg received a copy of a message from General Howard to General Meade informing the latter that a large cavalry force was seen marching towards the right flank of the Union army. At the same time General Gregg also received an order from General Pleasanton to send Custer's brigade to Kilpatrick as originally ordered.

To comply with the latter order McIntosh's brigade was directed to relieve Custer, which it at once proceeded to do.

McIntosh replaced Custer's squadrons in the woods at the cross roads by the 1st New Jersey, and placed his other two regiments, the 1st Maryland and 3rd Pennsylvania, south of or in rear of these woods. Custer assembled his brigade and prepared to move off the field.

CONFEDERATE CAVALRY.—Stuart's cavalry reached the vicinity of Gettysburg on the morning of July 3, where he was joined by Jenkins' brigade. He spent some hours in resting his horses and men, in issuing ammunition, etc.; then, with the brigades of Chambliss and Jenkins, he moved over to the York road and along the intersecting road towards the Low Dutch road. He knew from his map that the road he was following would enable him to reach the Baltimore road, only four miles distant, at a point two miles east of Rock Creek. He would thus be in rear of the enemy and on his line of communications: the very point from which he could make a diversion in favor of General Lee. As he reached the bend of the intersecting road at the Stallsmith farm, he halted his column behind the woods and rode into the field to scan the country. He saw that the country was open and sloped gradually southward. None of the Union cavalry was in sight; Gregg's brigade was concealed by the woods and ridge on Stuart's right, McIntosh by the woods at the forks of the road, and Custer by distance and minor obstructions on the field. Jenkins and Chambliss were placed in the woods and a messenger was sent to bring up Lee and Hampton.

THE BATTLE FIELD.—The field upon which Stuart's cavalry met that of Gregg is about three miles east of Rock Creek and lies in the northwest angle of the *Hanover* and *Low*

Dutch roads. Its western boundary is *Cress' ridge*, which extends from *Granite Hill* on the York road to the Hanover road at Cress' house; its elevation, where it crosses the road by which Stuart advanced, is about one hundred feet higher than its elevation at the Hanover road or at the Hanover-Low Dutch crossroads.

The field is traversed by the intersecting road upon which Stuart advanced and by a private road further south running east and west near the *Rummel* house. The east branch of Cress' Run, called *Little's Run*, rises near this road and flows parallel to and a mile west of the Low Dutch road. A wood, about four hundred yards long and two hundred yards wide, extends from the intersecting road on which the Confederates reached the field, southward along the ridge. This is the wood in which the Confederates were concealed. Another wood about four hundred yards square was in the southwest angle of the Low Dutch road and its intersecting road. In these woods rested the right flank of the Union line. Otherwise the ground was open, but divided into fields by stone and rail fences, which had to be thrown down for mounted maneuvers.

The *Lott* house is a half mile from the Hanover road and about three hundred yards west of the Low Dutch road. The Rummel house with its large frame barn is about a mile from the Hanover road and about three hundred yards south of the woods in which the Confederates were concealed.

THE CAVALRY FIGHT.—About 2 P.M., immediately after reaching the angle of the road, Jenkins' brigade took possession of the Rummel place and Chambliss was put in the woods on his left. When McIntosh observed the troops at the Rummel place he sent the 1st New Jersey dismounted to develop their strength. Jenkins met this movement, which he soon observed, by sending his dismounted troops to occupy the line of fence along Little's Run. McIntosh sent a part of the 3rd Pennsylvania to reinforce the 1st New Jersey and the remainder to occupy the woods vacated by the 1st New Jersey. He also sent word to General D. McM. Gregg that he could not alone hold the enemy with his three regiments. The latter thereupon ordered General Irvin Gregg to move to the Low Dutch road, and ordered Custer to remain and support McIntosh until further orders. Upon reaching the field itself and seeing Lee and Hampton appear, he ordered McIntosh's third regiment, the 1st Maryland, to occupy the woods in its front and protect his right flank.

Along Little's Run the Confederate line was strengthened by Chambliss and the line was prolonged to the left by skirmishers sent to the front by Hampton and Lee when they reached the field. The attacking line was strengthened on the left by the 5th Michigan and on the right by a squadron of the 6th Michigan. The attack was aided by Pennington's and Randol's batteries near the Hanover-Low Dutch cross roads, which soon silenced the Confederate battery which appeared in front of the woods north of Rummel's.

At length Jenkins' troops, who had only ten rounds of ammunition, were compelled to retire and the Union line advanced to the creek. Shortly thereafter the Union line retired for the same reason and was pursued by two of Chambliss' regiments, the 9th and 13th Va., one mounted and the other dismounted. This was the beginning of the mounted action.

The 7th Michigan mounted was sent to protect the right flank of the retreating troops and encountered Chambliss' regiment in the northwest corner of the field containing the Lott house; there the two regiments fought over a stone and rail fence. While thus engaged, the 1st Virginia, one of General Fitz Lee's regiments, charged down on the right of the 7th Michigan and two of Hampton's regiments, the 1st N. C. and the Jeff. Davis' legion, on its front and left. The 7th Michigan fell back and the 1st Virginia almost reached Chester's platoon of artillery in rear of the Lott house, when it was forced to retire. Hampton's regiments were met by the fire of Pennington's artillery and were charged in flank by the 5th Michigan. They were also forced back.

This was followed by the final charge of the remaining regiments of Fitz Lee's and Hampton's brigades, save a regiment of each held in reserve. This attack, made in column of squadrons, was met first by the artillery and then by the last regiment of Custer's brigade, the 1st Michigan, under his personal leadership. The column was charged in flank by the squadrons of the 1st New Jersey and 3rd Pennsylvania, which had fought dismounted in the earlier part of the engagement and were now again mounted; and also by such scattered troopers as could be collected by General McIntosh himself. These successive shocks caused the Confederates to fall back to their original positions and at 5 P.M. the fight was over. In this last attack General Hampton was seriously wounded.

Immediately after the fight the Union picket line was

re-established along the line of Little's Run, which had been the scene of the dismounted fighting, and the Confederate line was established at Rummel's and in the woods to the north.

Irvin Gregg's brigade was not engaged in the fight, but was held in reserve along the Low Dutch road south of the Hanover road.

That night Stuart retired to the west of Gettysburg and Custer joined Kilpatrick.

CAVALRY OPERATIONS ON THE UNION LEFT FLANK.

On the morning of July 3, General Kilpatrick with Farnsworth's brigade of cavalry moved from Two Taverns to the fields southwest of Big Round Top and being joined by Merritt's brigade of Buford's division formed a line extending from Plum Run to Willoughby Run. The right of the line rested on a wooded knoll, covered with boulders, which is separated from Big Round Top by Plum Run gorge. Farnsworth's brigade held this knoll and a line extending a short distance to the left; Merritt prolonged his line to Willoughby Run. One battery of artillery was on the wooded knoll and another near the Emmitsburg road.

To meet this flanking movement, General Law, who now commanded Hood's division, sent the 1st Texas of Robertson's brigade to form a line across the valley from Plum Run to Snyder's Ridge; this regiment intrenched itself behind the stone fences and boulders at the foot of the wooded knoll occupied by Farnsworth. Anderson's brigade was also withdrawn from the front line and sent to prolong the line of the

1st Texas to Willoughby Run. The brigades of Robertson and Law lay behind stone walls, which they had thrown up on the face of Big Round Top parallel to Plum Run. Hood's artillery was along Snyder's Ridge behind this line.

The only operations of note on this flank were the mounted charges made by different units of Farnsworth's brigade, in the last of which the brigade commander was killed.

The field covered by these charges was limited on the south by the wooded knoll, on the east by the Confederate intrenched line along Big Round Top, on the north by the stone fence below Devil's Den and on the west by the Confederate line on Snyder's Ridge. East of Plum Run the slopes of Big Round Top are wooded and covered with boulders, with the exception of a small cleared field on the wood road connecting Big Round Top with the Emmitsburg road. West of Plum Run the field is open, but divided into several small tracts by stone and rail fences. It was also swept by the fire of the batteries on Snyder's Ridge. In this open field are two houses with outbuildings.

In all the charges the mounted troops were formed to the left and rear of the wooded knoll and moved to its left.

About 5 P.M. General Kilpatrick decided to strike in rear the brigades of Law and Robertson and thus make a diversion in favor of the Union infantry in their front. He first ordered the 1st West Virginia to charge the 1st Texas infantry and clear the way. Farnsworth, with the two squadrons of the 1st Vermont, was then to charge Law and Robertson.

The 1st West Virginia charged the 1st Texas and pierced

its lines, but as the Confederate regiment was protected by walls and fences little damage was done. Immediately thereafter the first squadron (four troops) of the 1st Vermont rode along the north slope of the wooded knoll and straight across the fields to the wood road north of the house near Plum Run. At the road the squadron turned to the right and crossed the run, where it was re-formed near the cleared field. The third squadron led by Farnsworth in person, followed the first, but skirted the north slope of the wooded knoll, crossed Plum Run and rode along the Confederate line on the slope of Big Round Top as far as the wall south of Devil's Den; here the squadron divided, some of the troops riding straight across the open fields to the point of starting, while Farnsworth, with others, rode back along the Confederate line to the cleared field near which the first squadron was rallied; here he was killed. The first squadron charged through the cleared field and then turned to the right and rode through the gorge between Big Round Top and the wooded knoll.

Two of the Confederate regiments, one of Law's regiments east of Plum Run and one of Anderson's west of Plum Run, left their trenches and moved to the edge of the woods to take the charging columns in flank.

No movement was made by the Union infantry on Big Round Top during these charges, because the field of operations was not visible from the infantry line.

CASUALTIES IN BATTLE OF GETTYSBURG.

The losses in the three days' battle are as follows:

(W. R. No. 43, p. 173.)

	Killed.	Wounded.	Missing.	Total
I corps	666	3231	2162	6059
II "	797	3194	378	4369
III "	593	3029	589	4211
V "	365	1611	211	2187
VI "	27	185	30	242
XI "	369	1922	1510	3801
XII "	204	812	66	1082
Cavalry "	91	354	407	852
Artillery reserve	43	187	12	242
Totals	3155	14525	5365	23045

(W. R. No. 44, p. 338.)

	Killed.	Wounded.	Missing.	Total
Longstreet's corps	910	4339	2290	7539
Ewell's "	809	3823	1305	5937
Hill's "	837	4407	1491	6735
Cavalry "	36	140	64	240
Totals	2592	12709	5150	20451

The number of prisoners taken, wounded or unwounded in the field by the Union army was..........12,227

The greatest numerical losses by brigades in the *Union army* were:

Meredith	I corps,	1153	Carr	III corps,	790
Paul	I "	1041	Ward	III "	781
Cutler	I "	1002	Brewster	III "	778
Rowley	I "	898	Ames	XI "	778
Stone	I "	853	Harrow	II "	768
Schimmelfennig	XI "	807	Willard	II "	714

The greatest numerical losses by brigades in the *Confederate army* were:

Armistead,	Longstreet's corps,	1191	Iverson	Ewell's corps,	820
Pettigrew	Hill's "	1105	Wilcox	Hill's "	777
Garnett	Longstreet's "	941	Barksdale,	Longstreet's "	747
Daniel	Ewell's "	916	Kemper,	Longstreet's "	731

CASUALTIES AMONG PROMINENT OFFICERS.

Killed or Mortally Wounded.

UNION.		CONFEDERATE.	
Maj. Gen'l	Reynolds	Maj. Gen'l	Pender
Brig. "	Farnsworth	Brig. "	Armistead
Brig. "	Weed	Brig. "	Barksdale
Brig. "	Zook	Brig. "	Garnett
Colonel	Vincent	Brig. "	Semmes
Colonel	Willard	Colonel	Avery
Colonel	Cross		

Wounded.

Maj. Gen'l	Hancock	Maj. Gen'l	Hood
Maj. "	Sickles	Maj. "	Pickett
Brig. "	Gibbon	Maj. "	Heth
Brig. "	Paul	Maj. "	Trimble
Brig. "	Graham	Brig. "	Anderson
Brig. "	Barlow	Brig. "	Kemper
Colonel	Stone	Brig. "	Law
Colonel	Biddle	Brig. "	Jones
Colonel	Brooke	Brig. "	Scales
		Brig. "	Pettigrew
		Brig. "	Hampton

Captured.

Brig. Gen'l	Barlow	Maj. Gen'l	Trimble
Brig. "	Graham	Brig. "	Archer
Brig. "	Schimmelfennig		

All of the above officers commanded corps, divisions or brigades at the beginning of the battle.

Brig. Gen'l Pettigrew was killed at Williamsport in the retreat.

CHAPTER VIII.

THE RETREAT AND PURSUIT.

On the night of July 3-4, Ewell's corps was withdrawn to Seminary Ridge and Stuart was ordered to move to the extreme right of the Confederate army, leaving Fitz Lee's and Hampton's brigades on the Cashtown road to assist in guarding the trains. General Imboden was ordered to collect, on the Cashtown road, on the following morning, all the ammunition, artillery and general supply trains which were not to accompany the troops, all the wounded and prisoners, and escort them to Williamsport through Cashtown. The regiments left by Early in Virginia were ordered to Williamsport, and a supply of ammunition was ordered to the same place from the depot at Winchester.

On the same night Buford, who had just reached Westminster, was ordered to march to Frederick with the brigades of Gamble and Devin, and Huey's brigade of Gregg's division was ordered from Manchester to report to Kilpatrick at Emmitsburg.

On July 4 General Lee gave his final orders for the retreat. The army was to move through Fairfield, Monterey Pass and Hagerstown to Williamsport, near which the Confederates had a pontoon bridge, and the river was usually fordable in summer. The trains, protected by Iverson's brigade, were to move first, and then the corps of Hill, Longstreet, and Ewell. The movement was to begin at dark.

Robertson, who was at Fairfield, with his two cavalry brigades, was to assist in protecting the trains, while Stuart, with the brigades of Chambliss and Jenkins, was to move to Emmitsburg, and watch the roads leading to the mountains from that town.

THE CAVALRY AND CONFEDERATE TRAINS.—At 4 P.M. July 4, in a severe rain storm, Imboden started with his trains for Williamsport, marching through Cashtown, Greenwood and Greencastle. His column, which was seventeen miles long, was escorted by his own brigade, assisted by those of Fitz Lee and Hampton and twenty-three guns. He was pursued by Irvin Gregg's cavalry brigade. The head of his column reached Williamsport on the afternoon of the 5th, and on the morning of the 6th all the wagons were at the river. The two cavalry brigades, which acted as a rear guard, did not reach Williamsport until that evening. At Greenwood, Irvin Gregg overtook the rear guard and had an unimportant skirmish with it. At Greencastle, the column was attacked by a picked body of 200 cavalry sent from Milroy's old command, which had been assembled near Bedford, Pennsylvania; they captured or destroyed about sixty wagons before they could be driven off.

When Imboden reached Williamsport, he found that the pontoon bridge at Falling Waters, upon which Lee had relied for his crossing, had been destroyed by a cavalry force sent from Frederick the day before, and that the heavy rains of the 4th had raised the river above its fording level. He was therefore obliged to intrench his position and send his wagons over on two small flat boats. He was here reinforced by two of Early's regiments from Virginia, which escorted an ammunition train from Winchester.

On the afternoon of July 6, Imboden was attacked by Buford's entire division on the Boonsborough-Williamsport road and by Custer's brigade on the Hagerstown-Williamsport road. Kilpatrick had left the brigades of Huey and Farnsworth to protect his rear against Stuart's cavalry. By arming the teamsters and the least seriously wounded soldiers, Imboden was able to hold his own until the approach of Stuart from Hagerstown caused Buford to fall back and guard his own communications.

The trains were not again attacked; before the army crossed the river, Imboden's column had been ferried over and sent to Winchester. (Battles and Leaders, Vol. III, page 420.)

At 7 P.M. July 4 the army corps trains began to move along the Fairfield road towards Monterey Pass. The train was not properly guarded, since Iverson's brigade had not reached its position in the column, and Stuart had not had time to place himself astride the road running from Emmitsburg to Monterey Pass. This road was guarded only by a single company of Confederate cavalry on patrol duty.

The train having been observed by the signal officers on Big Round Top, Kilpatrick, who had moved to Emmitsburg in the morning, was directed to move on Monterey Pass and attack it. Kilpatrick reached the mountains about dusk in a rain storm. The officer in command of the Confederate patrol at once rode to the train passing in his rear and ordered the wagons to be driven in haste to Williamsport. He also ordered the wagons which had not reached the fork of the road to be stopped. He then returned to his station and made such demonstrations of resistance that he deceived Kilpatrick and succeeded in delaying his advance. He thus saved the first section of the train, which reached Williamsport in safety. As the road had to be cleared for the troops, the head of the second section of the train could stop only long enough to close up and was then obliged to move on. When Kilpatrick finally brushed aside the patrol, he found a part of the second section of the train in the road in his front; he took it with him to Smithsburg, a town across the mountains from Mechanicstown.

On the morning of July 5, Stuart, with the brigades of Chambliss and Jenkins, reached Emmitsburg and there learned that he was too late to stop Kilpatrick. He at once decided to seek him by crossing the mountains by the Mechanicstown road. As he was leaving the mountains in the afternoon, he found Kilpatrick in his front, and was unable to get through the passes. Towards evening, however, Kilpatrick withdrew to Boonsborough and Stuart advanced to Smithsburg.

On the morning of July 6, Buford and Kilpatrick with six brigades of cavalry were at Boonsborough and it was agreed between them that Kilpatrick with his three brigades should go to Hagerstown and delay the advance of the Confederate army, while Buford marched to Williamsport to capture the trains. Stuart was at Smithsburg on the morning of July 6, where he was joined by the brigades of Robertson and Jones. That morning he set out for Hagerstown and found Farnsworth's brigade of Kilpatrick's division in possession of the town and Huey's brigade in support. Believing this to be a covering force and that the main force was attacking the trains at Williamsport, Stuart attacked vigorously with all his cavalry, aided by Iverson's brigade of infantry. He drove Farnsworth's brigade back on the Boonsborough road and Huey's on the Williamsport road. It was the noise of Stuart's guns in their rear that caused Buford and Custer to retire from Imboden's front before piercing his lines. That night Buford and Kilpatrick were again in Boonsborough, and Stuart's cavalry, now joined by the brigades of Fitz Lee and Hampton, was posted along the Antietam.

On the morning of July 7, Stuart strengthened his line by securing Wofford's infantry brigade from Longstreet; this was posted on the road from Boonsborough to Williamsport in rear of the cavalry. Anderson's brigade of Hood's division was at this time on the Antietam at Funkstown.

On July 9 and 10 there were also indecisive engagements on the Boonsborough-Hagerstown roads in which Anderson's ~~division~~ brigade of Longstreet's corps took part.

On the 10th the infantry intrenched line was established between Hagerstown and the Potomac river and the cavalry was withdrawn from the front and placed on the left flank.

THE INFANTRY.—The Confederate infantry started for Williamsport on the night of July 4 and on the night of July 5 Hill bivouacked at Waynesborough, Longstreet in Monterey Pass, and Ewell at Fairfield. Ewell's rear guard was attacked by the VI corps.

The night of July 6, Longstreet was two miles beyond Hagerstown, Hill at Hagerstown, and Ewell beyond Monterey Pass.

On July 7 Ewell reached Hagerstown.

As soon as General Meade learned that the Confederate army was withdrawing from Gettysburg, he decided that a direct pursuit through the passes of the South mountains would be impracticable; if he found that General Lee was really retreating, he proposed to move the Union army parallel to the Hagerstown road, through the passes west of Frederick. He therefore ordered General French, who was stationed at Frederick with some of the troops withdrawn from Harper's Ferry, to occupy these passes at once.

July 5 was spent by General Meade in ascertaining the position and intentions of the Confederate army as well as in the care of the wounded and the dead of both armies. Irvin Gregg's brigade on the right reconnoitered as far as Hunterstown; the XI corps advanced one brigade into Gettysburg; the VI corps pursued the retreating Confederate army as far as Fairfield.

From the information obtained on this day, General Meade decided that the Confederate army was really retreating to the Potomac instead of simply retiring to the passes of the South mountains. He therefore ordered all his corps to march for the passes west of Frederick, leaving only one brigade of the VI corps, with McIntosh's cavalry brigade, to follow the Confederate army along the Fairfield road.

On July 9 the V and VI corps were across the mountains at Boonsborough, and the II and XII corps were at Rohrersville at the western base of the mountains due east of Shepherdstown; all the other corps were in close supporting distance. As the bridge over the Potomac at Falling Waters was still unfinished and the river unfordable, upon the approach of these corps, the Confederate army deployed on a line extending from the Potomac river a short distance below Falling Waters northward to a point one mile due west of Hagerstown. This line was thoroughly intrenched and well defended by artillery; Longstreet held the right, Hill, the center, and Ewell, the left. The cavalry was in rear of Ewell.

On July 12, the entire Union army was deployed and intrenched on a line extending from Funkstown, on the Antietam, to a point on the Hagerstown-Sharpsburg road opposite the Confederate right flank. The right wing was composed of the XI, I and VI corps, the left wing of the V, II and XII corps; the III corps was in reserve. Kilpatrick's cavalry was on the right flank, Buford's on the left, and Gregg in reserve. Since reaching the river the army had been reinforced by the four brigades of French's division.

The Union army also had a reserve composed of three brigades of New York and Pennsylvania militia, which had been assembled by General Couch for the defense of Harrisburg. On the retreat of the Confederate army, these brigades took the field as a division, commanded by General W. F. Smith. On July 12 he was a short distance northeast of Hagerstown.

On the night of July 12-13, General Meade called a council of war and submitted to his corps commanders the question of the advisability of assaulting the Confederate lines on the following day, July 13. The commanders in favor of attacking were Generals Wadsworth and Howard; those opposed, were Generals Hays, French, Sykes, Sedgwick and Slocum. It was therefore decided to defer the assault.

On July 13 the bridge at Falling Waters was finished and the river at Williamsport reached the fording stage.

General Lee now decided to retire into Virginia, as it was difficult to subsist his army, and he feared the Union army would be greatly reinforced before making an assault. He directed the infantry and artillery to withdraw from the trenches immediately after dark and the cavalry to replace them. Longstreet was ordered to cross at once over the bridge at Falling Waters, five miles in his rear, and Hill and the cavalry along his front were ordered to follow him. Ewell and the rest of the cavalry were directed to ford the river at Williamsport.

The 13th of July was overcast and misty and the night following dark and stormy. This made it easy to withdraw the Confederate army without attracting the attention of the Union troops, but made it difficult for the troops to move rapidly.

At 5 P.M. the artillery was started to the rear and immediately after dark Longstreet's infantry followed. During the night the bridge was broken by the heavily loaded wagons and for two hours communication was interrupted; it was therefore 9 A.M. on the 14th before Longstreet's corps was across. He was at once followed by two squadrons of cavalry which had been left to cover Hill's rear, but by mistake crossed the river ahead of him.

Ewell's corps moved to Williamsport after dark and there forded the river after midnight. The artillery ammunition chests and one brigade were taken across in boats. The crossing was completed at 8 A.M. on the 14th. With the exception of the two cavalry squadrons mentioned above, Stuart's cavalry crossed at Williamsport with Ewell.

On the morning of July 14, Kilpatrick moved forward

and finding the Confederate lines evacuated pushed on to Williamsport only to find Ewell and Stuart on the Virginia side. He then moved down to Falling Waters.

Hill's divisions marched all night in the rain in the order Anderson, Pender, Heth. In the morning Heth posted his division in line on the hills about two miles from the bridge to check the pursuers, while the artillery and trains were crossing. Pender's division was his reserve. About 11 A.M., after the artillery, trains, and Anderson's divisions had crossed, Heth was ordered to send Pender's division to the bridge. He was also directed to hold his line with one brigade, and to direct the others to follow Pender's division.

About this time Kilpatrick appeared in his front. An advance guard of Custer's brigade consisting of about fifty mounted troopers at once charged the Confederate line. This movement deceived the Confederates, who thought this body must be a troop of their own rear guard cavalry. The Confederate infantry therefore delayed firing upon them until the cavalry was almost upon the Confederate line. The Confederates had only two men killed by this charge, but one of them was General Pettigrew; two-thirds of the Union cavalry were killed. This was followed by an advance of one of Custer's regiments dismounted; he was driven back by Brockenbrough, whose brigade thus became detached from the division. At this moment Buford appeared on the field and moved his dismounted cavalry to turn Heth's right flank and intercept his retreat. Heth was about to call up Pender's brigades to meet this movement when he was ordered to retreat at once.

He sent one brigade to form a line a quarter of a mile

nearer the bridge, and then retired his other brigades through it; by repeating this operation he reached the bridge. Brockenbrough's brigade was too much involved to withdraw easily when the order was given, and therefore left on the field three regimental colors and a number of prisoners. Hill's corps was also obliged to abandon two guns and a large number of stragglers who had been overcome by the fatigue of the night march in a rainstorm. It is probable, however, that the total losses on this day did not exceed one thousand.

At 1 P.M., the Confederate army being on the Virginia shore, the bridge was cut loose, and General Lee's offensive campaign was over.

CHAPTER IX.

COMMENT ON GETTYSBURG.

CONFEDERATE OPERATIONS.

Though the battle of Chancellorsville, fought in the early days of May, 1863, was a Confederate victory, it did not materially improve the condition of affairs in Virginia. A week later the armies of the Potomac and of Northern Virginia were in the same relative positions they had occupied before the battle.

The Army of Northern Virginia was rendered more confident by its success in that campaign, and its faith in its commander was even stronger than ever: The army had, however, suffered an irreparable loss in the death of its most brilliant corps commander, General Thomas J. Jackson.

While the Army of the Potomac had lost some of its confidence in the ability of its commander, it had by no means lost confidence in its own fighting powers. The corps that had been but little engaged were eager to meet their adversaries at once. Before resuming the offensive, however, it was necessary to remount the cavalry, and to reorganize the infantry brigades that were daily losing regiments whose terms of service had expired.

General Lee was no doubt extremely gratified by the outcome of the Chancellorsville campaign, yet he was well aware that the victory had been won by taking such chances as he would be unwilling to risk again. When he divided his army in the presence of overwhelming odds, he ran great risk of being destroyed in detail. His success was due largely to the genius of General Jackson and to the errors of his adversary. Now that Jackson was no more and his adversary had learned the true use of cavalry, it was improbable that there would be a second Chancellorsville surprise.

He felt confident that if General Hooker was left undisturbed on the north bank of the Rappahannock, the regiments he was daily losing would soon be replaced and when his army was thoroughly reorganized he would again assume the offensive. Profiting by his numerical strength and the errors of his previous campaign, it was fair to assume that this time he would be successful and force Lee to retire to the North Anna.

General Lee was too great a commander to willingly assume a passive role, when he had an opportunity of striking a blow. He knew that ultimate success must go to him who can assume and maintain the offensive. He determined therefore to recruit his army to its maximum possible strength at once and open the new campaign before General Hooker had had time to reorganize. He withdrew the divisions of Hood and Pickett from southern Virginia and ordered them to rejoin the army. He reinforced his infantry by the brigades of Pettigrew and Davis from southern Virginia and North Carolina, as well as by two regiments from western Virginia. His cavalry he reinforced by the brigade of Robertson from North Carolina and that of Jenkins from western Virginia. General Lee was able to withdraw troops from southern Virginia because at this time the IX corps was ordered from southern Virginia to Kentucky.

While Lee's plan of operations was somewhat like that

of Hooker in his Chancellorsville campaign, it was designed to be much more decisive in its results. Hooker planned only to force Lee to retire from the line of the Rappahannock; the latter proposed to compel Hooker not only to give up the line of the Rappahannock, but to retire behind the Potomac. General Lee's plan was to accomplish this result, if possible, without a battle and to fight only at some point north of the Potomac river. Without wholly uncovering the main road to Richmond, he proposed to assemble his army at Culpeper and from that point to hurl against Winchester a force which should not only capture the division which Milroy had incautiously assembled there, but should in addition so threaten the northern states that the authorities would hasten to recall Hooker from the Rappahannock and direct him to defend the capital. The Army of Northern Virginia could then be assembled on the Potomac, march into Pennsylvania and draw Hooker far north of his base. At some point in that state it was hoped that Hooker could be decisively defeated in battle. Secrecy and celerity were necessary to the accomplishment of this plan.

The reorganization of the Army of Northern Virginia was completed towards the end of May, 1863, but Lee's advance was delayed a few days by the sudden appearance of a Union force in the vicinity of White House, on the Pamunkey river. To ascertain its meaning, both Pickett and Pettigrew were halted, on their march northward, in the vicinity of Hanover Junction. Being convinced that the movement was not a serious one, on June 2 Pickett and Pettigrew were ordered to join their corps and the campaign began.

Ewell and Longstreet withdrew quietly from the trenches

behind Fredericksburg and moved to Culpeper Court House, where Stuart, with five of his cavalry brigades, was awaiting them. Hill's corps was left in the trenches to hold the direct line to Richmond, until Hooker retired from his front. From these instructions, it is evident that General Lee did not expect Hill to be seriously attacked. He hoped that Hooker would remain in ignorance of his movements until Winchester was assailed and captured; he was satisfied that this attack would cause Hooker to receive orders to retire at once to protect Washington.

On the evening of June 9th, General Lee knew that the presence of Ewell's infantry at Brandy Station would lead General Hooker to suspect that the Confederate army was divided and might be defeated in detail either by an attack on Fredericksburg or an attack on Culpeper. It required courage, therefore, to adhere to his original plan and on the morning of the 10th send Ewell's corps to Winchester. Winchester is fifty miles from Culpeper, and the latter is thirty miles from Fredericksburg. General Lee was undoubtedly much relieved, when on June 12th he learned from his outposts that the Union army had abandoned the Rappahannock and was moving northward. His general plan was now successfully inaugurated.

Ewell's operations at Winchester were not so successful as they should have been, due to his failure to entirely surround that town. Had Rodes moved from Berryville towards Winchester and cut off communications between that town and the Potomac, Milroy and his whole command would have been captured. Ewell's orders to Rodes may have been in compliance with instructions to appear on the Potomac as soon as possible, and may have been simply due to an error of judgment. Rodes, himself, should have moved on Winchester without orders, since he knew that General Lee was anxious to capture Milroy.

The movement of the Confederate army from the Rappahannock to the Potomac, with its marching flank covered by Stuart's cavalry, was an extremely well conducted march. Having a cavalry screen between them and the enemy and the possession of the roads on both sides of the Blue Ridge mountains, the infantry made the march in a surprisingly short time.

The first and the most serious error in the conduct of the campaign, was made when General Stuart was authorized to pass around the Union army and cross the Potomac river between that army and Washington. This error seems to have arisen as follows: From his "War Reminiscences," we learn that Colonel Mosby, the famous partisan leader of Virginia, was near Middleburg, Virginia, June 10, 1863, organizing the first company of his battalion. On that day he crossed the Bull Run mountains and marched across country for Seneca Falls on the Potomac, where the following morning he surprised a body of Union troops and took many prisoners. He then returned to Middleburg and there awaited General Stuart, to whom on June 17th he described his exploit. Impressed by Mosby's story, it is probable that sometime between the 17th and 22nd, Stuart suggested to General Lee that the cavalry cross the Potomac at the same point.

On the morning of June 22, General Stuart reported by letter to General Lee that the Union troops were retiring to Aldie Gap. As this indicated that Hooker had no intention of advancing in force across the Loudon Valley, it pointed to a movement towards the Potomac and it was high time that Lee moved into Maryland. On receipt of Stuart's note, General Lee directed Longstreet, whose headquarters were near his own, to withdraw his troops from the Blue Ridge mountains and prepare to follow the other corps. Recurring to the suggestion of General Stuart, he discussed the matter with General Longstreet and finally decided to let Longstreet decide the point of crossing when he was in a position to dispense with the protection afforded by Stuart. Both generals interpreted Stuart's suggestion as a proposition to pass around the rear of the Union army.

Longstreet, in "Manassas to Appomattox," says that he, Longstreet, was to decide whether Stuart was to cross the Potomac above or below the Union army.

As a result of that interview, there followed this correspondence:

HEADQUARTERS, June 22nd, 1863.

MAJ. GEN. J. E. B. STUART,
Commanding Cavalry.

General, I have just received your note of 7.45 this morning to General Longstreet. I judge the efforts of the enemy yesterday were to arrest our progress and ascertain our whereabouts. Perhaps he is satisfied. Do you know where he is and what he is doing? I fear he will steal a march on us, and get across the Potomac before we are aware. If you find that he is moving northward, and that two brigades can guard the Blue Ridge and take care of your rear, you can move with the other three into Maryland, and take position on General Ewell's right, place yourself in communication with him, guard his flank, keep him informed of the enemy's movements, and collect all the supplies you can for the use of the army. One column of General Ewell's army will probably move toward the Susquehanna by the Emmitsburg route; another by Chambersburg. Accounts from him last night state that there was no enemy west of Frederick. A cavalry force (about 100) guarded the Monocacy Bridge, which was barricaded. You will, of course, take charge of (A. G.) Jenkins' brigade, and give him necessary instructions. All supplies taken in Maryland must be by authorized staff officers for their respective departments—by no one else. They will be paid for, or receipts for the same given to the owners. I send you a general order on this subject, which I wish you to see is strictly complied with.

I am, very respectfully, your obedient servant,

R. E. Lee,

General.

(W. R. No. 45, p. 913.)

Headquarters,

June 22, 1863—7.30 p.m.

General R. E. Lee,

Commanding, and so forth.

General: Yours of 4 o'clock this afternoon is received. I have forwarded your letter to General Stuart, with the suggestion that he pass by the enemy's rear if he thinks that he may get through. We have nothing of the enemy to-day.

Most respectfully,

JAMES LONGSTREET,

Lieutenant-General, Commanding.

(W. R. No. 45, p. 915.)

HEADQUARTERS,

Millwood, June 22, 1863—7 P.M.

MAJ. GEN. J. E. B. STUART,

Commanding Cavalry.

General Lee has inclosed to me this letter for you (of same date), to be forwarded to you, provided you can be spared from my front, and provided that I think that you can move across the Potomac without disclosing our plans. He speaks of your leaving, via Hopewell Gap, and passing by the rear of the enemy. If you can get through by that route, I think that you will be less likely to indicate what our plans are than if you should cross by passing to our rear. I forward the letter of instructions with these suggestions.

Please advise me of the condition of affairs before you leave, and order General Hampton—whom I suppose you will leave here in command—to report to me at Millwood, either by letter or in person, as may be most agreeable to him.

Most respectfully,

JAMES LONGSTREET,

Lieutenant-General.

N. B.—I think that your passage of the Potomac by our rear at the present moment will, in a measure, disclose our plans. You had better not leave us, therefore, unless you can take the proposed route in rear of the enemy. (W. R. No. 45, p. 915.)

As General Lee's letter made no mention of Stuart's proposed plan approved by Longstreet, but laid great stress on the necessity of his reaching the Potomac as soon as possible, Stuart was somewhat in doubt as to the best route to follow. Colonel Mosby, being in Stuart's camp that night, the latter sent for him to get his views. Mosby urged that the shortest practicable line to the Potomac was the line Gum Springs-Seneca Falls, *through* the Union army. The Union corps being widely separated, he contended that the movement could be easily made. En route much damage could be done to the Union trains and the communications between the Union army and the capital could be severed. Mosby says this was the plan Stuart adopted, and that he was to meet Stuart at Gum Springs, to conduct his advance guard. After the interview, General Stuart sent a messenger to General Lee and the following night received this reply:

HEADQUARTERS, ARMY OF NORTHERN VIRGINIA,
June 23, 1863—5 P.M.

MAJ. GEN. J. E. B. STUART,
Commanding Cavalry.

General: Your notes of 9 and 10.30 A.M. to-day have just been received. As regards to the purchase of tobacco for your men, supposing that Confederate money will not be taken, I am willing for your commissaries or quartermasters to purchase this tobacco and let the men get it from them, but I can have nothing seized by the men.

If General Hooker's army remains inactive, you can leave two brigades to watch him, and withdraw with the three others, but should he not appear to be moving northward, I think you had better withdraw to this side of the mountain to-morrow night, cross at Shepherdstown next day, and moveover Fredericktown.

You will, however, be able to judge whether you can pass around their army without hindrance, doing them all the damage you can, and cross the river east of the mountains. In either case, after crossing the river, you must move on and feel the right of Ewell's troops, collect information, provisions, and so forth.

Give instructions to the commander of the brigades left behind, to watch the flank and rear of the army, and (in the event of the enemy leaving their front), retire from the mountains west of the Shenandoah, leaving sufficient pickets to guard the passes, and bringing everything clean along the valley, closing up on the rear of the army.

As regards the movements of the two brigades of the enemy moving toward Warrenton, the commander of the brigades to be left in the mountains must do what he can to counteract them, but I think that the sooner that you cross into Maryland, after to-morrow, the better.

The movements of Ewell's corps are as stated in my former letter. Hill's first division will reach the

Potomac to-day, and Longstreet will follow to-morrow. Be watchful and circumspect in all your movements.

I am, very respectfully and truly, yours

R. E. Lee,

General.

(W. R. No. 45, p. 923.)

It is difficult to understand General Stuart's reasons for his movement. On June 22nd he knew that General Lee considered it of the greatest importance that he should place himself on Ewell's right flank. There were three courses open to him to accomplish this result. The first was to go down the Loudoun valley and either cross the river east of Harper's Ferry, or traverse the mountains near Charlestown and the Potomac west of Harper's Ferry. The time necessary to make this movement could have been accurately calculated, as the routes were practically unobstructed. The second, was to ride *around* the Union army. As he had not ascertained the position of the rear corps of that army, he could not tell in advance how long his route would be or where he could cross the Potomac. It was certain to be a plunge in the dark. The third was to ride *through* the Union army, as advised by Colonel Mosby. This is probably what he intended to do, as it was the shortest line to the Potomac. This was certainly a hazardous operation with as large a force as a cavalry division. It is probable that he was fortunate in finding his route obstructed before he had penetrated the Union lines. Mosby, who went to Gum Springs to meet him, found himself in the midst of the Union army and found it difficult to escape.

While General Lee no doubt deeply regretted that he

had acceded to Stuart's plan, he can hardly be blamed for authorizing a plan approved by Generals Longstreet and Stuart, especially after he himself had so clearly indicated to the latter that haste was necessary and he must select a route over which he could march without delay.

The result of Stuart's action was far reaching. From the time Lee crossed the river into hostile territory, he had no reliable information of the movements of the Union army. The two brigades left by Stuart in the mountains, were practically useless in the absence of their division commander. On the night of June 30th, when they should have been with the advance guard at Cashtown, Penn., they were at Martinsburg, Virginia, two days in rear.

When General Ewell was directed to cross the Potomac river he was instructed to inform the inhabitants that his movement to Hagerstown was for the purpose of enveloping Harper's Ferry. By this ruse General Lee hoped to have this garrison withdrawn, as it menaced his communications.

It is not altogether clear what General Lee expected to accomplish in Pennsylvania besides drawing Hooker north of the Potomac and gathering supplies for his army. On June 22nd he wrote to General Ewell as follows:

HEADQUARTERS, June 22, 1863.

LIEUT. GEN. R. S. EWELL.

General: Your letter of 6 P.M. yesterday has been received. If you are ready to move, you can do so. I think your best course will be toward the Susquehanna, taking the routes by Emmitsburg, Chambersburg and McConnellsburg. Your trains had better be, as far as possible, kept on the center route. You must get command of your cavalry, and use it in gathering supplies, obtaining information, and protecting your flanks. If necessary send a staff officer to remain with General Jenkins. It will depend upon the quality of supplies obtained in that country whether the rest of the army can follow. There may be enough for your command, but none for the others. Every exertion should, therefore, be made to locate and secure them. Beef we can drive with us, but bread we cannot carry, and must secure it in the country.

I send you copies of a general order on this subject, which I think is based on rectitude and sound policy, and the spirit of which I wish you to see enforced in your command. I am much gratified at the success which has attended your movements, and feel assured, if they are conducted with the same energy and circumspection, it will continue. Your progress and direction will, of course, depend upon the development of circumstances. If Harrisburg comes within your means, capture it. General A. P. Hill arrived yesterday in the vicinity of Berryville. I shall move him on to-day if possible. Saturday, Longstreet withdrew from the Blue Ridge. Yesterday the enemy pressed our cavalry so hard with infantry and cavalry on the Upperville road that McLaws' division had to be sent back to hold Ashby's Gap. I have not yet heard from there this morning. General Stuart could not ascertain whether it was intended for a real advance toward the Valley or to ascertain our position.

The pontoons will reach Martinsburg to-day, and will be laid at the point you suggest, 4 or 5 miles below Williamsport, if found suitable. I have not countermanded your order withdrawing the cavalry from Charlestown. I will write you again if I receive information affecting your movements.

Trusting in the guidance of a merciful God, and invoking His protection for your corps, I am, with great respect, your obedient servant,

R. E. LEE,

General.

(W. R. No. 45, p. 914.)

From this it would appear that Ewell's movement was intended mainly to gather supplies for the army in and near the Cumberland Valley. He was directed to take Harrisburg, if it proved to be a feasible task. General Lee's movement against Hooker was to be planned later, after he learned from the cavalry the exact position of Hooker's army. From Longstreet's writings it appears that Lee hoped to maneuver so as to place himself between the Union army and Washington, and thus compel Hooker to attack him in position. In his own report, General Lee says that Early was sent to York to cause the Union army to drift in that direction, away from his own line of communications.

He was without information of the position of Hooker's army from the morning of June 22, when Stuart reported Pleasanton retreating, until June 28, when he learned from Longstreet's spy that the Union army was around Frederick. He then felt it incumbent upon himself to cross the mountains to draw Hooker away from Frederick. Had he known that the Union army was under orders to move towards York on the morning of the 29th, this movement would have been unnecessary. The Union army was now moving in the direction he had designed it should move when Early was sent to York. His natural maneuver was not to interfere with that movement, but to concentrate quietly and move towards Washington by Fairfield and Emmitsburg, or by some other route even farther south. He would then place himself on the flank or in rear of the Union army, with his communications secure and could either attack his enemy in flank or rear, or await attack in a favorable position. Longstreet's spy left the Union army a few hours too soon to bring General Lee this valuable information.

Although the movement of General Lee's army to Gettysburg was forced upon him, it produced a condition of affairs which might have been fatal to the Union army had General Lee known the true situation of that army.

On the afternoon of July 1, the I and XI Union corps, the smallest of the army, found themselves in a very critical position. South of the Cashtown road, in the fields west of Seminary Ridge, were the three brigades of Stone, Meredith and Rowley. Their only supports were the artillery of the I corps on the ridge in their rear, and the cavalry brigade of Gamble to their left and rear in the valley behind Seminary Ridge. In their front were the eight brigades of the divisions of Heth and Pender of Hill's corps.

At the north end of Seminary Ridge, widely separated from its left wing, were the three brigades of Baxter, Paul and Cutler forming the right wing of the I corps. In their front were the brigades of O'Neal, Iverson, Daniel and Ramseur of Rodes' division of Ewell's corps.

In the low ground, north of the town, far in rear of the right flank of the I corps, were four brigades of the XI corps. On their right and rear was Devin's cavalry brigade; in their front was Doles' brigade of Rodes' division, and coming down the Heidlersburg road to turn their right, were the four brigades of Early's division.

The general reserve of the whole line consisted of two brigades of the XI corps, which were with the commander of the field, General Howard, on Cemetery Hill.

All that was necessary to give the Confederates a decisive victory was a commander on the field who could see the entire situation. While Rodes was fighting a delaying action with the troops in his front, a vigorous attack could have been made on Rowley, Meredith and Stone and at the same time their left flank could have been turned. In this attack the whole of Hill's force, eight brigades, could have been employed, since Anderson's division was close behind. Had these troops attacked, as they and others did on the 2nd and 3rd of July, they could hardly have failed to destroy the three small brigades in their front. Having captured the ridge at the Seminary, they would have been in a position to cut off the retreat of the right wing of the I corps and the advanced brigades of the XI corps. The position of the latter could, in the meantime, have been turned by Early. As the Union troops had but a single line of retreat, that up Cemetery Hill, it is fair to presume that the two corps would have been almost annihilated by the superiority of position and numbers of the Confederate forces engaged. If these two corps could have been destroyed, General Lee woul have been in an excellent condition to meet the remainin corps of the Union army.

The failure of the Confederates to profit by their advantages on this day may be attributed to a single cause, defective information. General Lee did not expect a battle this day and did not reach the field until the afternoon fight was begun, too late to change any of the dispositions for attack even had he desired to do so. West of the town, General Hill, who commanded in the critical part of the field, was unaware of the importance of Seminary Hill, since the ridge concealed from him the town and the position of the XI corps. He therefore made very defective tactical dispositions for its capture. So badly were these made in fact that instead of engaging with his eight brigades the entire attention of Doubleday's three, one of these latter, Stone's brigade, had time to turn its attention to Rodes' division and make a flank attack on Daniel's brigade. The result was that Seminary Hill was not captured until the rest of the Union troops were re-forming on Cemetery Hill. Had Hill carried Seminary Hill early in the afternoon, General Lee would have had an opportunity of directing operations from that point, where he would have seen the precarious position of the other Union troops on the line.

When General Lee reached Seminary Ridge, the battle seems to have been practically over. Hill's troops were resting on the ridge. Ewell's men were in the town, and the Union troops were re-forming on Cemetery Hill. Lee saw at a glance that the battle was not won unless Cemetery Hill was also taken, and he at once sent word to Ewell to take it if he felt able to do so.

Ewell reported that his troops were worn out with marching and fighting and that he was unable to bring his artillery to bear on the Union line. He would therefore postpone the attack until Johnson's division arrived. It is possible that Ewell was wrong in making this decision, but he was probably right. It would have taken him some time to re-form his troops, who must have been in considerable confusion; he would then have been obliged to advance without artillery support against a strong position. It is possible that before he was ready to move, the XII corps would have reached the field. The opportunity of taking the hill was lost when the Union troops were not vigorously pursued in their retreat by all the troops of both Ewell and Hill.

The opportunity of fighting a battle under favorable conditions, one of the objects for which he had moved north of the Potomac, came to General Lee on July 1, but he was ignorant of that fact until it was too late. He was destined not to have another.

After the battle of July 1, from the cupola of the college, General Lee carefully examined the Union position upon Cemetery Hill. He then decided that it would be too strong to assault on the morrow. He therefore directed General Ewell to move to the west of the town and take his position on the right of Hill's corps. From the report of General Pendleton, his chief of artillery, who had made an examination of the ground along this front, he decided that an attack could be made from this direction. This attack he hoped to make, with the divisions of McLaws, Hood, Anderson and Johnson, before the Union army was united; all of these divisions were arriving on the field and had not been previously engaged.

The order for the movement of Ewell's corps was, however, countermanded at the request of the corps commander. Early in the evening of July 1, two of Ewell's staff officers had ascended Culp's Hill and found it unoccupied. Having reported this at headquarters, Ewell asked permission to occupy this hill early in the morning with Johnson's division and thus compel the Union troops to evacuate Cemetery Hill. He would thus accomplish what General Lee had requested him to accomplish by assault the night before. This plan would have been eminently successful had the Union army not been marching that night for Gettysburg, and had not the commander on Cemetery Hill realized the necessity of occupying Culp's Hill. During the night Johnson learned from an intercepted messenger that Sykes' V corps was coming along the Hanover road, and from a reconnoitering party that Culp's Hill was occupied by some Union infantry. The information thus received deterred him from acting without further orders. Before the wishes of Generals Ewell and Lee could be ascertained, morning broke and it was too late to move Ewell's corps, as first proposed. He was therefore left in a position from which he was obliged to attack without artillery support, and with his wings separated by the town, so that concerted action along his line was impossible. His partial success, on July 2, was due wholly to an error made by the Union commander, and not to any virtue in his own position.

Would the attack on the Union left wing have been successful on July 2, had Ewell moved as General Lee first directed? This is extremely doubtful, for the Union right flank would then have rested on Cemetery Hill and the XII and V corps would have extended the line to the Round Tops. General Meade would have given more attention to the left of the line, which could have been made safe.

It would, however, have placed General Lee in a position to shift his line of communication to the Emmitsburg road and maneuver about the left flank of the Union army and thus compel Meade to evacuate the lines of Gettysburg. This would have been the great advantage secured by the shifting of Ewell's corps.

It has been said that the battle of July 2 was not a more decided success because of the failure of Longstreet to attack at daylight. That General Lee desired to attack as soon as possible on the 2nd, is only natural, for he knew that the other Union corps must be marching towards Gettysburg. That he expected an attack at daylight seems hardly plausible. On July 1, only a single officer of his army, General Pendleton, had examined this part of the field and he was hampered by the presence of Buford's cavalry. Before any disposition for attack could be made, it would be necessary for General Lee and his staff to examine the ground. Time was also necessary to move the troops, who were encamped along the Cashtown road, into position. The evidence of the records is clear that General Anderson, who was to connect Longstreet and Pender and support Longstreet's attack, was not in position until noon. As no criticism has ever been made of Anderson's movements, it would appear that he moved as rapidly as was expected, and the attack could not have been made before noon. It was delayed from noon until 4 P.M. by the officers sent from headquarters to guide Longstreet's divisions from the Cashtown road to the ~~left~~ right of Anderson's position. Being more or less ignorant of the topography of the country, they delayed the movement in their endeavor to reach the position without exposing the troops to the Union signal station on Little Round Top. Had Longstreet been able to attack an hour or two earlier, he would have found the III corps changing position and less able to resist his assaults than at 4 P.M.

When General Lee decided to renew the battle on July 3, he probably assumed that a combined effort of Generals Pickett and Johnson would pierce the Union line, weakened by the demoralization caused by the Confederate victories of July 1 and 2. He probably also counted on Stuart to assist in this effort. Had he been able to view the situation as it really was, he would not have ordered the attack, since Johnson was certain to be defeated. It is difficult to understand why the divisions of Pickett and Heth were allowed to advance after Johnson's defeat. In addition to advancing over open fields exposed to artillery fire, when they reached their goal, they were almost certain to meet a vastly superior force of infantry.

The position occupied by the Confederate army on July 2 and 3 was undoubtedly abad one. With fewer troops than General Meade, General Lee occupied a longer line, which was nearly everywhere commanded by the Union position and was interrupted by the town of Gettysburg. It offered no good points from which an attack could be made.

Such success as Lee had was due almost wholly to the absence of cavalry on the Union left flank on July 2, which enabled Longstreet and Anderson to concentrate in front of that flank without the knowledge of General Meade. The whole situation on the left would have been changed, had Meade known in the morning that Anderson and Longstreet were advancing towards his left.

It is difficult to imagine a more discouraging situation than that in which General Lee found himself between the 4th and 14th of July. Decisively repulsed in battle and compelled to retreat, his communications were suddenly severed by the destruction of his only bridge and by the floods at the fords. Yet it is clear, that never once through these trying days, did the commander or his men show any sign of demoralization. On the contrary, it is certain, they would have welcomed an attack on their intrenched lines about Falling Waters.

Union Operations.

To attack the fraction of the Confederate army in his front, as soon as he ascertained that the Confederate army was divided, was a most natural plan for General Hooker to suggest to the President, and it was equally natural for General Halleck to reply that if either fraction was to be attacked, it should be the one which crossed the Rappahannock rather than the one left in the trenches at Fredericksburg. General Hooker knew where he would find Hill's corps, but the corps moving towards Culpeper were not so easily brought to battle. The President and General Halleck on the other hand remembered the depression throughout the North produced by Lee's invasion of Maryland and Jackson's Shenandoah Valley operations, and could not consent to a movement which did not conform to Hooker's general orders of "covering Washington and Harper's Ferry."

Neither plan was wrong, but of the two that of General Halleck was the safer. Hill might have declined to await Hooker in his trenches and marched at once to join Lee by the shortest practicable route. With the Confederate army on the Potomac, a movement on Richmond, as General Hooker proposed, was one to which the Northern people would never have given their consent.

The retrograde movement of General Hooker from the Rappahannock to Frederick was well executed in all its details. The excellent employment of cavalry is especially to be commended.

When he reached Frederick, General Hooker was naturally much embarrassed. His orders were to cover Washington and Harper's Ferry and also to seek the Confederate army and attack it. The Confederate army, being reported at York, Carlisle and Chambersburg, it was manifestly impossible to carry out all these instructions, and as he was not allowed to abandon Harper's Ferry and absorb its garrison into his army, he submitted his resignation.

When relieved, General Hooker was planning a movement northward with his flanks on opposite sides of the South Mountains. His left column was to be composed of the XII corps, strengthened by the Harper's Ferry garrison, and was to move up the Cumberland valley on Lee's line of communication. As Lee ordered his concentration east of the mountains and practically abandoned his communications, General Hooker would thus have run the risk of having his right wing destroyed before the left wing crossed the mountains. General Meade's plan of operations was a safer one, since there were no mountain passes between his wings.

If there was any error in General Meade's advance from Frederick, it was in advancing with too great a front. It would have been better, from a tactical point of view, to move the VI corps to Hanover rather than to Manchester. It is possible, however, that this would have been too great a concentration for convenience of supply. That he realized the disadvantages of his extended front is evident from his selection of the Pipe Creek line, to which his forces could retire should either wing suddenly meet an overwhelming force. It was a wise provision, for which he has received more criticism than praise.

General Meade also realized that General Lee's most probable move would be to turn his left flank. He therefore placed three corps on this flank and combined them in a single command under his most trusted corps commander. He cautioned him to reconnoiter carefully the ground about Emmitsburg in view of a possible battle there.

General Meade intended to advance cautiously, for he realized that he was facing a commander of great genius, and that the result of the campaign would be of the utmost importance to the North and to the South. While he did not shun a battle he did not care to engage in one in which General Lee had any advantage, through position or numbers.

In spite of all his precautions and through no apparent

error of any commander, his left wing was drawn into the battle of July 1, in which the chances were all against it. General Buford reached Gettysburg about noon of June 30 and at once sent patrols along all the roads, to ascertain the position of the Confederate troops. He reported late that night that he believed all of Hill's corps was at Cashtown, that Rodes was reported coming from Carlisle, and that rumor said that some troops were coming from York. There was nothing in the information received by General Reynolds from General Buford that should have led Reynolds to act otherwise than he did. He moved on Gettysburg in accordance with his previous orders, taking the precaution to have the XI corps close at hand.

The position of the Union army was not an unfavorable one, while Heth and Rodes were alone on the field; but when Pender and Early appeared on the two flanks the position of the Union forces was rendered extremely hazardous. At 1 o'clock, as soon as General Howard knew that Rodes was in the vicinity, he sent word to the commanders of the III and XII corps that he was engaged. These corps were, however, unable to reach the field until the battle was over. Had he known that Pender and Early were so near, he would probably have withdrawn the whole command to Cemetery Hill before the afternoon engagement. Had he sent word to General Slocum as soon as he reached the field in the morning, the latter would have been on the field in the early afternoon. The head of Slocum's corps was at Two Taverns, five miles from Gettysburg, at 11 A. M.

When General Meade sent Hancock to inspect the field of Gettysburg, he did not know that the III and XII corps had been notified and were on their way to assist General Howard. He probably thought that the main battle was to be fought on Pipe Creek and that it was advisable for him to remain where he had established his headquarters for the day. Had he known that these corps were moving towards Gettysburg, he would probably have gone there in person. In the course of the afternoon, however, it became evident that the battle must be fought at Gettysburg, and he accordingly sent his orders to the V and VI corps and moved up to the field.

When General Meade arrived on the field of Gettysburg, he was impressed by its defects. It was a position which could easily be turned, as General Hancock had said in his report, and presented a very narrow front to the troops of Ewell and Hill whom he saw on the north, stretched from Seminary Ridge on the west to Benner's Hill on the east. Johnson was encamped along the Hanover road in close proximity to his main line of communications, the Baltimore pike. He naturally wanted to drive Johnson from this position, and early in the morning of July 2 directed that an attack be made by the XII and V corps as soon as the latter should come up. It was because of this proposed movement that he failed to critically examine the position of his left wing.

There were two serious errors made on July 2 which contributed greatly to such success as the Confederates won—the removal of Buford's cavalry from the field, and the withdrawal of most of the XII corps from its trenches.

General Pleasanton seems to have had the same idea as General Meade, that the left flank of the Union army would be safe as soon as the III corps reached the field. He therefore ordered Buford to move off the field and go to Westminster to guard the trains. Had Buford been on the field, all the movements of Anderson and Longstreet would have been discovered as soon as made. Had Longstreet's movement been known in time, the left wing of the Union army could have been strengthened to resist it, by the use of the V corps.

When Barksdale, Wilcox, Perry and Wright's brigades followed the retreating line of the III corps about 6 P.M., two divisions of the V and one of the II had already been thrown into line to support the III Corps. As reserves, General Meade had but one division of the V corps left. As the VI corps was just coming up and he did not know to what extent he could rely upon it for assistance, he sent for the XII corps. As it turned out this additional force was unnecessary and was not employed. The absence of the XII corps from its trenches enabled Johnson to get a foothold on the slopes of Culp's Hill. Only the stubborn resistance of Greene's brigade, which was retained at General Slocum's request, prevented Johnson from securing the position which had been Ewell's objective since the preceding evening.

Although the movement of General Sickles' corps to its position in advance of the general line was not contemplated by General Meade, it was probably a fortunate move, as it gave the warning which should have been given by the cavalry and broke up Longstreet's attack before it reached the main line. As the foreground, over which Longstreet advanced against Little Round Top, was broken and wooded, it would have been difficult, if not absolutely impossible, for the III corps alone to have stopped Longstreet's corps, had it reached Devil's Den and the ground in its vicinity, intact. The ground was quite different from that over which the charge was made the following day and was not easily swept by either infantry or artillery fire.

The disposition of the Union army on July 3 was an ideal one for passive defense. Its flanks were well protected by cavalry, its infantry and artillery were well distributed, and stone walls were thrown up wherever the material on the site permitted it; as on the Round Tops and Culp's Hill. Not knowing at what point General Lee would launch his next attack, General Meade distributed the VI corps along the line wherever he thought it needed strengthening, and therefore had no organized reserve ready to follow the retreating troops after Pickett's charge. To have moved forward with his main line without previous preparation, and have led to confusion, and the advance would have probably been stopped by the line of Longstreet's guns, even though the Confederate batteries had exhausted much of their ammunition.

The operations of the cavalry on the right flank of the Union army on July 1, 2 and 3 were extremely well conducted. It was fortunate that Custer made the mistake of moving to the right flank of the army on the 3rd, instead of going to the left flank where he was expected.

The operations on the left flank were not so successful. On the 2nd Buford received the unfortunate orders which led to his absence at a critical time; and on the 3rd Kilpatrick made an ill-judged attack which caused the loss of many of his men without the possibility of its affecting the Confederate infantry concealed by woods and stone fences.

General Meade has been much criticised for allowing the Confederate army to cross the Potomac without vigorously attacking it. This criticism does not seem to be just, for the following reasons:

General Lee had the shortest line to Williamsport and Falling Waters, and had it not been for the flood in the river and the destruction of his pontoon bridge, would have been across the river long before General Meade could have overtaken him. It would have been impossible for any force to effect much on his direct line of retreat, since his line lay through a mountain pass where the pursuer could be readily stopped. General Meade made undoubtedly the right move when he launched his cavalry in pursuit and followed with his infantry via Frederick and the passes nearer the river than those followed by General Lee. The records show that the cavalry began to move for Frederick on July 4, and on the same day the infantry moved from Frederick to seize the passes through the mountains. The movement of the infantry from the battlefield began the following day.

On the evening of July 12 the Union army, reinforced by the Harper's Ferry garrison, was in front of the Confederate intrenched line, which extended from the vicinity of Hagerstown to the river. That night General Meade called a meeting of his corps commanders and requested their opinions as to the advisability of an attack. Of the seven corps commanders, five considered that the chances of success were not sufficient to warrant the risk of losing the moral and material effect of Gettysburg. They believed that the main object of the campaign was to drive the Confederate army out of the Northern states, and that that object should be gained without running any unnecessary risk. Without an assault it was certain that Lee would cross the river; if the assault failed, he would probably renew his invasion. Under the circumstances, General Meade could hardly be blamed for accepting the judgment of such generals as Slocum, Sedgwick and Sykes.

Those who at the time criticised General Meade's action, assumed that the Confederate army was demoralized and out of ammunition, and that an assault must have resulted in the annihilation of the Confederate army. We now know that the Army of Northern Virginia was well supplied with ammunition, and instead of dreading an assault, was eager to have the opportunity of repulsing one. In view of Spottsylvania and Cold Harbor, it is probable that the assault would have been repulsed. If the assault had been repulsed, the losses would have been great, and this would have done much to neutralize the moral effects of Gettysburg and Vicksburg.

MEADE, GEORGE GORDON. Born in Spain, 1815. Graduate U. S. M. A., 1835. Second lieutenant of artillery, 1835-1836. Resigned. Re-entered service as lieutenant of topographical engineers. First lieutenant, 1851; captain, 1856; major, 1862; brigadier general, 1863; major general, 1864. Brevetted first lieutenant for conduct in battle of Monterey, Mexican war. Brigadier general of volunteers, August 31, 1861; major general of volunteers, November 29, 1862-1864. On surveys and engineering duties, 1836-1861. At battles of Palo Alto, Resaca de la Palma, and Monterey, and at the siege of Vera Cruz in Mexican war, 1846-1847. Commanded brigade in Peninsula campaign, taking part in battles of Mechanicsville, Gaines' Mills, and Glendale, and later in battle of Manassas. Commander of a division, September to December, 1862. Engaged in battles of South Mountain and Antietam. Corps commander, December, 1862-June, 1863. Engaged in battles of Fredericksburg and Chancellorsville. Commander of Army of the Potomac, 1863-1865. Engaged in all the operations of that army beginning with Gettysburg, 1863, and terminating with Appomattox, 1865. After the war in command of various military departments and divisions. Died 1872.

REYNOLDS, JOHN F. Born in Pennsylvania, 1820. Graduate U. S. M. A., 1841. Second lieutenant of artillery, 1841; first lieutenant, 1846; captain, 1855; lieutenant colonel of infantry, 1861; colonel of infantry, 1863. Brevetted captain for conduct in battle of Monterey, and captain for conduct in battle of Buena Vista, in Mexican War. Brigadier general of volunteers, August 20, 1861; major general of volunteers, November 29, 1862. On artillery duty, 1841-1861. At defence of Fort Brown and in battles of Monterery and Buena Vista in Mexican war. Commanded brigade in Peninsula campaign of 1862, being engaged in battles of Mechanicsville, Gaines' Mills, and Glendale. Commander of division in battle of Manassas, and of Pennsylvania state militia during Antietam campaign. Corps commander, November, 1862-July, 1863, and as such engaged in battles of Fredericksburg, Chancellorsville, and Gettysburg. Killed at Gettysburg, July 1, 1863.

HANCOCK, WINFIELD S. Born in Pennsylvania, 1824. Graduate U. S. M. A., 1844; second lieutenant of infantry, 1844; first lieutenant, 1853; captain, quartermaster's department, 1855; major, 1863; brigadier general, 1864; major general, 1866. Brevetted first lieutenant for conduct in battles of Contreras and Cherubusco, Mexican War. Brevetted major general, U. S. A., for conduct in battle of Spottsylvania. Brigadier general of volunteers, September 23, 1861; major general of volunteers, November 29, 1862. On infantry duty, 1844-1855. In engagements at Plan del Rio, San Antonio, Cherubusco, Molino del Rey, and City of Mexico, in Mexican War. On quartermaster duty on the frontier 1855-1861. As brigade commander in Peninsula campaign in engagements at Yorktown, Williamsburg, Golding's Farm, Savage Station, White Oak Swamp, and in Maryland campaign at Antietam. Division commander in the campaigns and battles of Fredericksburg and Chancellorsville. Corps commander and as such engaged in all the operations of the Army of the Potomac from Chancellorsville to Petersburg, except when temporarily absent because of wound received at Gettysburg. November, 1864, to February, 1865, organizing corps of veterans. After the war in command of various military departments and divisions. Died 1886.

SICKLES, DANIEL E. Born in New York City, 1825. Member of state and national legislature, 1857-1861. Colonel of regular infantry, 1866; retired with rank of major general, 1869. Colonel of New York volunteers, 1861; brigadier general of volunteers, 1861; major general of volunteers, 1863; mustered out, 1868. Brigade commander before Richmond, 1862; division commander at Fredericksburg; corps commander at Chancellorsville and Gettysburg. Lost a leg at Gettysburg. On special mission to South America in 1865. Since retirement has filled various legislative and executive offices.

SYKES, GEORGE. Born at Dover, Delaware, October 9, 1822. Graduate U. S. M. A., 1842; lieutenant of infantry, 1842-1855; major, 1855-1861; lieutenant colonel, 1863; colonel, 1868. Brigadier general of volunteers, September 28, 1861; major general of volunteers, November 29, 1862. Brevetted captain for conduct in battle of Cerro Gordo, Mex. Served in Florida (Indian wars); on frontier duty in the south and south-west; in Texas; in the Mexican war; and again on frontier. Served in Bull Run campaign; division commander in the Peninsula, at Manassas, Antietam, Fredericksburg, and Chancellorsville; corps commander at Gettysburg and in the subsequent campaigns of the Army of the Potomac. Later again on frontier and garrison duty. Died February 18, 1880.

SEDGWICK, JOHN. Born September 13, 1813, at Cornwall, Connecticut. Graduate U. S. M. A., 1837; artillery officer until 1855, when appointed a major of cavalry. Brigadier general of volunteers, August 31, 1861; major general, July 4, 1862. Brevetted captain for conduct in battles of Contreras and Cherubusco; major, for conduct in battle of Chapultepec, Mex. Served in the Florida war, on the western and Canadian frontiers, and throughout the Mexican War in the artillery; and in Indian wars in the cavalry. Division commander in the Peninsula compaign until wounded, June 30, 1862; rejoined and again wounded at Antietam; corps commander at Chancellorsville and Spottsylvania. Killed at Spottsylvania C. H., May 9, 1864.

HOWARD, OLIVER O. Born in Maine, November 8, 1830. Graduate U. S. M. A., 1854; lieutenant of ordnance, 1854-1861; resigned June 7, 1861; brigadier general, December 21, 1864; major general, March 19, 1886. Colonel of Maine Volunteers, June 4, 1861; brigad'er general of volunteers, September 3, 1861; major general of volunteers, November 29, 1862. Before the war served at various arsenals, in the Florida Indian troubles, and as instructor in mathematics at the U. S. M. A. Commanded a brigade at Bull Run, commanded XI Corps at Chancellorsville, Gettysburg and in Sherman's advance on Atlanta. Became commander of the Army of the Tennessee, July, 1864, and served as such in the operations of Sherman's army to the close of the war. After the war commanded various territorial departments and divisions and served a tour of duty as superintendent U. S. M. A.

SLOCUM, HENRY W. Born in New York. Graduate U. S. M. A., 1852; resigned as lieutenant of artillery, October 31, 1856; after volunteer service declined a regular commission as colonel. Colonel of New York volunteers, May 21, 1861; brigadier general of volunteers, August 9, 1861; major general of volunteers, July 4, 1862; resigned, September 16, 1865. Commanded his regiment at the first Bull Run; division commander in the Peninsula and at Antietam; corps commander at Chancellorsville, Gettysburg and thereafter. After the war returned to New York and re-entered public life, serving as congressman, etc., until his death in 1894.

PLEASANTON, ALFRED. Born in D. C. Graduate U. S. M. A., 1844; lieutenant of cavalry till March, 1855; captain of cavalry, 1855-1862; major of cavalry, 1862-1868, when he resigned after having declined a commission as lieutenant colonel of infantry; restored to the retired list by act of Congress in October, 1888. Brigadier general of volunteers, July 16, 1862; major general of volunteers, June 22, 1863. Brevetted first lieutenant for conduct in battles of Palo Alto and Resaca de la Palma, Mex. In his early career served on the frontier, in the Mexican war, and in various Indian expeditions; during the early days of the war was on recruiting and mustering duty; commanded a cavalry division at Fredericksburg and Chancellorsville and the cavalry corps thereafter until March, 1864, when he was sent to the west in charge of operations in Missouri and Kansas. After leaving the army occupied various public and private civil positions.

LEE, ROBERT E. Born in Virginia, 1807. Graduate U. S. M. A., 1829; engineer officer through grades to include captain until March 3, 1855, when appointed a lieutenant colonel of cavalry; colonel of cavalry, March 16, 1861; resigned, April 25, 1861. Brevetted major for conduct in battle of Cerro Gordo, Mex.; lieut.-colonel for conduct in battles of Contreras and Cherubusco; colonel for conduct in battle of Chapultepec, Mex. Brigadier general in the Confederate service, May 14, 1861; general, June 4, 1861; general-in-chief, January 31, 1865. As engineer officer in regular service was engaged in the civil duties of his corps, in the Mexican war, and as superintendent, U. S. M. A.; as cavalry officer in the regular service was engaged in frontier Indian duties, department commander, etc. As confederate officer was in command in West Virginia in the autumn of 1861; of the defences of the coast of South Carolina, Georgia, and Florida in the Spring of 1862; and after June 1, 1862, commanded the Army of Northern Virginia until the close of the war. After the war was President of Washington College, Lexington, Va., until his death in 1870.

LONGSTREET, JAMES. Born in South Carolina. Graduate U. S. M. A., 1842; infantry officer of grades to include captain, until July 19, 1858, when appointed major and paymaster; resigned June 1, 1861. Brevetted captain for conduct in battles of Contreras and Cherubusco; major, for conduct at battle of Chapultepec, Mex. Lieutenant colonel of confederate infantry, March 16, 1861; brigadier general, June 17, 1861; major general, October 7, 1861; lieutenant general, October 9, 1862. In U. S. service was engaged in frontier, garrison, Mexican war, and similar service. In confederate service, brigade commander at Bull Run; division commander on the Peninsula; commanded a wing, or corps at Manassas and thereafter throughout the operations of the Army of Northern Virginia until the end of 1863, when he was placed in command of the department of East Tennessee; later again a corps commander in the Army of Northern Virginia. After the war engaged in commercial pursuits and held various state and federal positions in Louisiana and Georgia until his death in 1904.

EWELL, R. S. Born in the District of Columbia. Graduate U. S. M. A., 1840; cavalry officer in various grades to include captain, until his resignation, May 7, 1861. Brevetted captain for conduct in battles of Contreras and Cherubusco, Mex. Lieutenant colonel of confederate cavalry, March 16, 1861; brigadier general, June 17, 1861; major general, January 24, 1862;

lieutenant general, May 23, 1863. In the federal service engaged in frontier, garrison, recruiting and other special and routine duties and in the Mexican war. In confederate service, brigade commander at Bull Run, division commander in Jackson's Shenandoah Valley campaign, and later around Richmond in the Peninsula campaign; also at the battle of Manassas, where he lost a leg; corps commander from Gettysburg until end of 1864, when assigned to the command of the Department of Richmond, which he retained until the end of the war. After the war was a farmer in Tennessee until his death in 1872.

HILL, A. P. Born in Virginia. Graduate U. S. M. A., 1847; served as lieutenant of artillery until his resignation on March 1, 1861. Colonel of confederate infantry, 1861; brigadier general, February 22, 1862; major general, May 26, 1862; lieutenant general, May 24, 1863. While in U. S. service, took part in Mexican war, and was on frontier and garrison duty and scientific duty in Washington. In the confederate service, regimental commander at Bull Run; brigade and division commander in the Peninsula; division commander at Manassas, Antietam, and Fredericksburg; division and corps commander at Chancellorsville, and corps commander thereafter until killed at Petersburg, April 2, 1865.

STUART, J. E. B. Born in Virginia. Graduate U. S. M. A., 1854; lieutenant and captain of cavalry until his resignation, May 14, 1861. Captain confederate cavalry, March 16, 1861 (later colonel); brigadier general, September 24, 1861; major general, July 25, 1862. On frontier, garrison and similar duties in Federal service. Commanded a regiment of cavalry at Bull Run; commanded the brigade of cavalry in the Peninsula in 1862, and entire cavalry, or the cavalry division of the Army of Northern Virginia in all of its subsequent operations until mortally wounded at Yellow Tavern, Va., May 10, 1864.

Army of the Potomac.

Commander-in-Chief—Major General George G. Meade (19-1835).
Chief of Staff—Major General Daniel Butterfield.
Chief of Artillery—Brig. General Henry J. Hunt (19-1839)
Chief of Engineers—Brig. General Gouverneur K. Warren (2-1850).

CORPS.	DIVISIONS.	BRIGADES.	REGIMENTS.
	Provost Guard.	Brig. General Marsena Patrick. (48-1835)	93d New York 8th U. S. Infantry (8 cos.)* 2d Pennsylvania Cavalry 6th Penn. Cavalry (2 cos.)
		Capt. Lemuel Norton. Guards and Ord'lies	Signal Corps Oneida (New York) Cavalry 1st Maine Cavalry (Co. L)
I. Major General Jno. F. Reynolds. (26-1841)	1st Brig. General James S. Wadsworth	1st Brig. General Solomon Meredith	19th Indiana 24th Michigan 2d Wisconsin 6th " 7th "
		2d Brig. General Lysander Cutler.	7th Indiana 76th New York 84th " (14th Militia 95th " 147th " 56th Pennsylvania (9 cos.)
	2d Brig. General John C. Robinson.	1st Brig. General Gabriel R. Paul. (18-1834)	16th Maine 13th Massachusetts 94th New York 104th " 107th Pennsylvania
		2d Brig. General Henry Baxter.	12th Massachusetts 83d New York (9th Militia 97th " 11th Pennsylvania 88th " 90th "
	3d Maj. General Abner Doubleday. (24-1842)	1st Brig. General Thomas A. Rowley.	80th New York (20th Militia) 121st Pennsylvania 142d " 151st "
		2d Colonel Roy Stone.	143d Pennsylvania 149th " 150th "
		3d Brig. General George J. Stannard. (Joined night of July 1st.)	12th Vermont (train guard) 13th " 14th " 15th " 16th "
		Artillery, Colonel Chas. S. Wainwright.	2d Maine Light Bat. (Hall) 5th " " 1st New York Light Battery 1st Pennsylvania Light Bat. 4th U. S. Battery B

*—Not engaged.

CORPS.	DIVISIONS.	BRIGADES.	REGIMENTS.
2d Major General Winfield S. Hancock. (8-1844)		General Headq'rs	6th New York Cavalry, Cos. D and K
	1st Brig. General John C. Caldwell.	1st Colonel Edward E. Cross.	5th New Hampshire 61st New York 81st Pennsylvania 148th "
		2d Colonel Patrick Kelly.	28th Massachusetts 63d New York (2 cos.) 69th " " 88th " " 116th Pennsylvania (4 cos.)
		3d Brig. General Samuel K. Zook.	52d New York 57th " 66th " 140th Pennsylvania
		4th Colonel John R. Brooke.	27th Connecticut (2 cos.) 2d Delaware 64th New York 53d Pennsylvania 145th " (7 cos.)
	2d Brig. General John Gibbon. (20-1847)	1st Brig. General William Harrow.	19th Maine 15th Massachusetts 1st Minnesota 82d New York (2d Militia)
		2d Brig. General Alexander S. Webb. (13-1854)	69th Pennsylvania 71st " 72d " 106th "
		3d Colonel Norman J. Hall. (13-1859)	19th Massachusetts 20th " 7th Michigan 42d New York 59th New York (4 cos.)
		Unattached.	1st Co. (Mass.) Sharpshooters
	3d Brig. General Alexander Hayes* (20-1844)	1st Colonel Samuel S. Carroll. (44-1856)	14th Indiana 4th Ohio 8th " 7th West Virginia
		2d Colonel Thomas A. Smyth.	14th Connecticut 1st Delaware 12th New Jersey 10th New York (Battalion) 108th "
		3d Colonel George L. Willard.	39th New York (4 cos.) 111th " 125th " 126th "
		Artillery, Captain John G. Hazard.	1st New York Light Bat. B. 1st Rhode Island Light Bat. A 1st Rhode Island Light Bat. B 1st U. S. Battery I 4th U. S. Battery A

CORPS.	DIVISIONS.	BRIGADES.	REGIMENTS.
V. Major. General George Sykes. (39-1842)		Provost Guard	12th New York, Cos. D&E 17th Penn. Cavalry, Co.D
	1st Brig. General James Barnes. (5-1829)	1st Colonel William S. Tilton.	18th Massachusetts 22d 1st Michigan 118th Pennsylvania
		2d Colonel Jacob B. Sweitzer.	9th Massachusetts 32d " 4th Michigan 62d Pennsylvania
		3d Colonel Strong Vincent.	20th Maine 16th Michigan 44th New York 83d Pennsylvania
	2d Brig. General Romeyn B. Ayres. (22-1847)	1st Colonel Hannibal Day. (22-1823)	3d U. S. (6 cos.) 4th " (4 ") 6th " (5 ") 12th " (8 ") 14th " (8 ")
		2d Colonel Sidney Burbank. (17-1829)	2d U. S. (6 cos.) 7th " (4 ") 10th " (3 ") 11th " (6 ") 17th " (7 ")
		3d Brig. General Stephen Weed. (27-1854)	140th New York 146th " 91st Pennsylvania 155th "
	3d Brig. General Samuel W. Crawford	1st Colonel William McCandless.	1st Pennsylvania Reserves 2nd " " 6th " " 13th " "
		3d Colonel Joseph W. Fisher.	5th Pennsylvania Reserves 9th " " 10th " " 11th " " 12th " (9 cos.)
		Artillery, Captain Augustus P. Martin.	3d Mass. Light Battery C 1st New York Light Battery 1st Ohio Battery L 5th U. S. Battery D 5th U. S. Battery I

CORPS.	DIVISIONS.	BRIGADES.	REGIMENTS.
III. Maj. General Daniel E. Sickles.	1st Maj. General David B. Birney.	1st Brig. General Charles K. Graham.	57th Pennsylvania (8 cos.) 63d " 68th " 105th " 114th " 141st "
		2d Brig. General J. H. Hobart Ward.	20th Indiana 3d Maine 4th " 86th New York 124th " 99th Pennsylvania 1st U.S. Sharpshooters (Berdan) 2d " " 8 cos.
		3d Colonel P. Regis de Trobriand	17th Maine 3d Michigan 5th " 40th New York 110th Pennsylvania (6 cos.)
	2d Brig. General Andrew A. Humphreys. (13-1831.)	1st Brig. General Joseph B. Carr.	1st Massachusetts 11th " 16th " 12th New Hampshire 11th New Jersey 26th Pennsylvania 84th " (train guard)
		2d Colonel Wm. R. Brewster.	70th New York 71st " 72d " 73d " 74th " 120th "
		3d Colonel George C. Burling.	2d New Hampshire 5th New Jersey 6th " 7th " 8th " 115th Pennsylvania
		Artillery, Captain George E. Randolph.	2d New Jersey Light Battery 1st New York Light Bat. D 4th " " 1st Rhode Island Light Bat. E 4th U. S. Battery K

CORPS.	DIVISIONS.	BRIGADES.	REGIMENTS.
VI. Maj. General John Sedgwick. (24-1837)		Headq'rs Guard	1st New Jersey Cavalry (1 co.) 1st Penn. Cavalry, Co. H
		Provost Guard	4th New Jersey (3 cos.)
	1st Brig. General Horatio G. Wright. (2-1841)	1st Brig. General A. T. A. Torbert. (21-1855)	1st New Jersey 2d " 3d " 15th
		2d Brig. General Joseph J. Bartlett.	5th Maine 121st New York 96th Pennsylvania
		3d Brig. General David A. Russell. (38-1845)	6th Maine 49th Pennsylvania (4 cos.) 119th Pennsylvania 5th Wisconsin
	2d Brig. General Albion P. Howe. (8-1841)	2d Colonel Lewis A. Grant.	2d Vermont 3d " 4th " 5th " 6th "
		3d Brig. General Thomas H. Neill. (27-1847)	7th Maine (6 cos.) 33d New York (detachment) 43d " 49th " 77th " 61st Pennsylvania
	3d Maj. General John Newton. (2-1842)	1st Brig. General Alexander Shaler.	65th New York 67th " 122d " 23d Pennsylvania 82d "
		2d Colonel Henry L. Eustis. (1-1842)	7th Massachusetts 10th " 37th " 2d Rhode Island
		3d Brig. General Frank Wheaton.	62d New York 93d Pennsylvania 98th " 102d " (train guard) 139th "
		Artillery, Colonel Chas. H. Tompkins.	1st Mass. Light Battery A 1st New York Light Battery 3d " " 1st Rhode Island Light Bat. C 1st Rhode Island Light Bat. G 2d U. S. Battery D 2d U. S. Battery G 5th U. S. Battery F

RPS.	DIVISIONS.	BRIGADES.	REGIMENTS.
Oliver O. Howard (4-1854)		Headq'rs Guard	1st Indiana Cav. (cos. I & K) 8th New York Infantry (1 co.)
	1st Brig. General Francis C. Barlow.	Colonel Leopold Von Gilsa.	41st New York (9 cos.) 54th " 68th " 153d Pennsylvania
		2d Brig. General Adelbert Ames. (5-1861)	17th Connecticut 25 Ohio 75th " 107th "
	2d Brig. General Adolph Von Steinwehr	Colonel Charles R. Coster.	134th New York 154th " 27th Pennsylvania 73d "
		2d Colonel Orland Smith.	33d Massachusetts 136th New York 55th Ohio 73d "
	3d Maj. General Carl Schurz.	1st Brig. General Alex. Schimmelfennig	82d Illinois 45th New York 157th " 61st Ohio 74th Pennsylvania
		2d Colonel W. Krzyzanowski.	58th New York 119th " 82d Ohio 75th Pennsylvania 29th Wisconsin
		Artillery, Major Thomas W. Osborn.	1st New York Light Bat. I 13th New York Light Battery 1st Ohio Light Battery K 1st Ohio Light Battery I 4th U. S. Battery G (Wilkeson)

CORPS.	DIVISIONS.	BRIGADES.	REGIMENTS.
XII. Maj. General Henry W. Slocum. (7-1852)		Headq'rs Guard	10th Maine (battalion)
	1st Brig. General Alpheus S. Williams.	1st Colonel Archibald L. McDougal	5th Connecticut 20th " 3d Maryland 123d New York 145th " 46th Pennsylvania
		2d Brig. General Henry H. Lockwood. (Joined July 2d) (22-1836)	g Maryland Potomac Home Maryland Eastern Shore 150th New York
		3d Brig. General Thomas H. Ruger. (3-1854)	27th Indiana 2d Massachusetts 13th New Jersey 107th New York 3d Wisconsin
	2d Brig. General John W. Geary.	1st Colonel Charles Candy.	5th Ohio 7th " 29th " 66th " 28th Pennsylvania 147th Pennsylvania (8 cos.)
		2d Colonel George A Cobham. Jr.	29th Pennsylvania 109th " 111th "
		3d Brig. General George S. Greene. (2-1823)	60th New York 78th " 102d " 137th " 149th "
		Artillery, Lieut. Edw. D. Muhlenberg.	1st New York Light Bat. M Pennsylvania Light Battery 4th U. S. Battery F 5th U. S. Battery K

CORPS.	DIVISIONS.	BRIGADES.	REGIMENTS.
Maj. General Alfred Pleasanton. (7-1844)	1st Brig. General John Buford. (16-1848)	1st Colonel William Gamble	8th Illinois 12th " (4 cos.) 3d Indiana (6 cos.) 8th New York
		2d Colonel Thomas C. Devin.	6th New York 9th " 17th Pennsylvania 3d West Virginia (2 cos.)
		Reserve, Brig. Gen'l Wesley Merritt. (22-1860)	6th Penn. 1st U. S. 2d " 5th " 6th "
	2d Brig. General David Mc M. Gregg. (8-1855)	Headq'rs Guard	1st Ohio, Co. A
		1st Colonel John B McIntosh.	1st Maryland (11 cos.) Purnell Legion, Md., Co. A 1st Massachusetts 1st New Jersey 1st Pennsylvania 3d " 3d " Heavy Artillery Section Battery H
		2d Colonel Pennock Huey.	2d New York 4th " 6th Ohio (10 cos.) 8th Pennsylvania (At Westminster during bat.)
		3d Colonel J Irvin Gregg.	1st Maine 10th New York 4th Pennsylvania 16th "
	3d Brig. General Judson Kilpatrick. (17-1861)	Headq'rs Guard	1st Ohio, Co. C
		1st Brig. General Elon J. Farnsworth.	5th New York 18th Pennsylvania 1st Vermont 1st West Virginia (10 cos.)
		2d Brig. General George A. Custer. (34-1861)	1st Michigan 5th " 6th " 7th " (10 cos)
	Horse Artillery.	1st Captain Jas. M. Robertson.	9th Michigan Battery 6th New York 2d U. S. Battery B, L and M 4th U. S. Battery E
		2d Captain John C. Tidball. (11-1848)	1st U. S. Batteries E, G & K 2d " " A (Cale. 3d " " C

CORPS.	DIVISIONS.	BRIGADES.	REGIMENTS.
		Headq'rs Guard	32d Massachusetts, Inf. 1
		1st Regular Capt. Dunbar R. Ransom.	1st U.S. Battery H 2d 2 batteries 4th Battery C 5th U.S. Battery C
		1st Volunteer Lt. Col. Freeman Mc Gilvery.	5th Mass. Light Battery E, 9th Mass. 15th NY Light Battery, Penn Light Battery, Penn Light Batteries (2)
	Artillery Reserve, Brig. General Robert O. Tyler (22-1853.)	2d Volunteer Capt. Elijah D. Taft.	1st Conn. Heavy Battery 1st Conn. H H eavy Bat Light Battery 5th New 1st 1st 1st New Hampshire Light
		3d Volunteer Capt. James F. Huntington.	Ohio Light Battery H Penn. Light Battery F & Virginia Light Battery
		4th Volunteer Capt. Robert H. Fitzhugh.	6th Maine Light Battery Maryland Light Battery A 1st New Jersey Light 1st N. Y. Light Battery
		Train Guard Maj. Charles Ewing.	4th N. J. Infantry (7 cos.)

* Not engaged.

Army of Northern Virginia.

Commander in Chief, General Robert E. Lee (2-1829.)
Chief of Artillery, Brigadier General William N. Pendleton (5-1830.)

CORPS.	DIVISIONS.	BRIGADES.	REGIMENTS.
Lieut. General James Longstreet. (54-1842)	Major General Lafayette McLaws. (48-1842)	Brig. General J. B. Kershaw	2d South Carolina 3d " 7tn " 8th " 15th " 3d South Carolina Battalion
		Brig. General Wm. Barksdale.	13th Mississippi 17th " 18th " 21st "
		Brig. General P. J. Semmes.	10th Georgia 50th " 51st " 53d "
		Brig. General W. T. Wofford.	16th Georgia 18th " 24th " Cobb's Legion, Ga. Phillip's Legion, Ga.
		Colonel H. C. Cabell.	Pulaski, Ga., Battery 1st N. Carolina Battery A Troup, Ga., Battery 1st Richmond Howitzers
	Major General George E. Pickett. (59-1846)	Brig. General R. B. Garnett. (29-1841)	8th Virginia 18th " 19th " 28th " 56th "
		Brig. General J. L. Kemper.	1st Virginia 3d " 7th " 11th " 24th "
		Brig. General L. A. Armistead.	9th Virginia 14th " 38th " 53rd " 57th "
		Major James Dearing.	Fauquier Battery Hampden Battery Richmond Fayette Battery Virginia Battery

CORPS.	DIVISIONS.	BRIGADES.	REGIMENTS.
Lieut.-General James Longstreet (54-1842)	Major General John B. Hood. (44-1853)	Brig. General E. M. Law.	4th Alabama 15th " 44th " 47th " 48th "
		Brig. General J. B. Robertson.	3d Arkansas 1st Texas 4th " 5th "
		Brig. General Geo. T. Anderson.	7th Georgia 8th " 9th " 11th " 59th "
		Brig. General H. L. Benning.	2d Georgia 15th " 17th " 20th "
		Major M. W. Henry.	Branch, N. C., Battery German, S. C., Battery Palmetto, S. C., Battery Rowan, N. C., Battery
	Corps Artillery, Colonel J. B. Walton.	Colonel E. P. Alexander. (3-1857)	Ashland, Va., Battery Bedford, Va., Battery Brooks, S. C., Battery Madison, La. Battery Virginia Battery Virginia Battery
		Major B. F. Eshleman.	Washington Artillery (4 batteries)

CORPS	DIVISIONS	BRIGADES	REGIMENTS
II. Lieut. General R. S. Ewell. (13-1840)	Major General Jubal A. Early. (18-1837)	Brig. General Harry T. Hays.	5th Louisiana 6th " 7th " 8th " 9th "
		Brig. General William Smith.	31st Virginia 49th " 52d "
		Hoke's Brigade Col. Isaac E. Avery.	6th N. Carolina 21st " 57th "
		Brig. General J. B. Gordon.	13th Georgia 26th " 31st " 38th " 60th " 61st "
		Lt. Colonel H. P. Jones.	Charlottesville Battery Courtney Battery Louisiana Battery Staunton Battery
	Major General Edward Johnson. (32-1838)	Brig. General George H. Steuart. (37-1848)	1st Maryland Battalion 1st N. Carolina 3d " 10th Virginia 23d " 37th "
		Brig. General James A. Walker.	2d Virginia 4th " 5th " 27th " 33d "
		Nichol's Brigade Col. J. M. Williams.	1st Louisiana 2d " 10th " 14th " 15th "
		Brig. General John M. Jones. (39-1841)	21st Virginia 25th " 42d " 44th " 48th " 50th "
		Major J. W. Latimer.	1st Maryland Battery Allegheny Battery Chesapeake Battery Lee Battery

CORPS	DIVISIONS	BRIGADES	REGIMENTS
Lieut. General R. S. Ewell (13-1840	Major General R. E. Rodes.	Brig. General Junius Daniel.	32d N. Carolina 43d " 45th " 53d " 2d " Battalion
		Brig. General Alfred Iverson.	5th N. Carolina 12th " 20th " 23d "
		Brig. General George Doles.	4th Georgia 12th " 21st " 44th "
		Brig. General S. D. Ramseur. (14-1860)	2d N. Carolina 4th " 14th " 30th "
		Colonel E. A. O'Neal.	3d Alabama 5th " 6th " 12th " 26th "
		Lieut. Colonel Thomas H. Carter.	Jefferson Davis Battery King William Battery Morris Battery Orange Battery
	Corps Artillery, Colonel J. T. Brown.	Captain Willis J. Dance.	2d Richmond Howitzers 3d " " Powhatan Battery Rockbridge Battery Salem Battery
		Lieut. Colonel William Nelson.	Amherst Battery Fluvanna Battery Georgia Battery

CORPS.	DIVISIONS.	BRIGADES.	REGIMENTS.
III. Lieut. General Ambrose P. Hill. (15-1847)	Major General R. H. Anderson. (40-1842)	Brig. General Cadmus M. Wilcox. (54-1846)	8th Alabama 9th " 10th " 11th " 14th "
		Brig. General A. R. Wright.	3d Georgia 22d " 48th " 2d " Battalion
		Brig. General William Mahone.	6th Virginia 12th " 16th " 41st " 61st "
		Perry's Brigade Col. David Lang.	2d Florida 5th " 8th "
		Brig. General Carnot Posey.	12th Mississippi 16th " 19th " 48th "
		Major John Lane	Co. A, Sumter Artillery Co. B " " Co. C " "
	Major General Henry Heth. (38-1847)	Brig. General J. J. Pettigrew.	11th N. Carolina 26th " 47th " 52d "
		Colonel J. M. Brockenbrough.	40th Virginia 47th " 55th " 22d " Battalion
		Brig. General James J. Archer.	5th Alabama Battalion 13th " 1st Tennessee 7th " 14th "
		Brig. General Joseph R. Davis.	2d Mississippi 11th " 42d " 55th N. Carolina
		Lieut. Colonel John J. Garnett.	Donaldsonville Battery Huger Battery Lewis Battery Norfolk Blues Battery

CORPS	DIVISIONS	BRIGADES	REGIMENTS
Lieut. General Ambrose P. Hill (15-1847)	Major General William D Pender. (19-1854)	Colonel Abner Perrin.	1st S. Carolina 12th " 13th " 14th " 1st " Rifles
		Brig. General James H. Lane.	7th N. Carolina 18th " 28th " 33d " 37th "
		Brig. General Edward L. Thomas.	14th Georgia 35th " 45th " 49th "
		Brig. General A. M. Scales.	13th N. Carolina 16th " 22d " 34th " 38th "
		Major William T. Pogue.	Albermarle Battery Charlotte Battery Madison Battery Virginia Battery
	Corps Artillery, Colonel R. L. Walker.	Major D. G. McIntosh.	Danville Battery Hardaway Battery 2d Rockbridge Battery Virginia Battery
		Major W. J. Pegram.	Crenshaw Battery Fredericksburg Battery Letcher Battery PeeDee Battery Purcell Battery

CORPS.	DIVISIONS.	BRIGADES.	REGIMENTS.
Cavalry.	Maj. General J. E. B. Stuart. (13-1854)	Brig. General Wade Hampton.	1st North Carolina 1st South Carolina 2d " " Cobb's Ga. Legion Jeff Davis Legion Phillips Ga. Legion
		Brig. General Beverly Robertson. (25-1849)	4th North Carolina 5th " "
		Brig. General Fitz Lee. (45-1856)	1st Maryland Battalion* 1st Virginia 2d " 3d " 4th " 5th "
		Brig. General A. G. Jenkins.	14th Virginia 16th " 17th " 34th Va. Battalion 36th " Jackson's Battery
		Brig. General William E. Jones. (10-1848)	6th Virginia 7th " 11th "
		Brig. General W. H. F. Lee. Colonel J. R. Chambliss, Jr. (31-1853)	2d N. Carolina 9th Virginia 10th " 13th "
		Horse Artillery Maj. R. F. Beckham. (6-1859)	Breathed's Va. Battery Chew's Va. Battery McGregor's Va. Battery Moorman's Va. Battery Griffin's Md. Battery Hart's S. C. Battery
	Independent Cavalry.	Brig. General J. D. Imboden.	18th Va. Cavalry 62d Va. Infantry (mtd.) Va. Partisan Rangers Virginia Battery

* With Ewell's Corps.

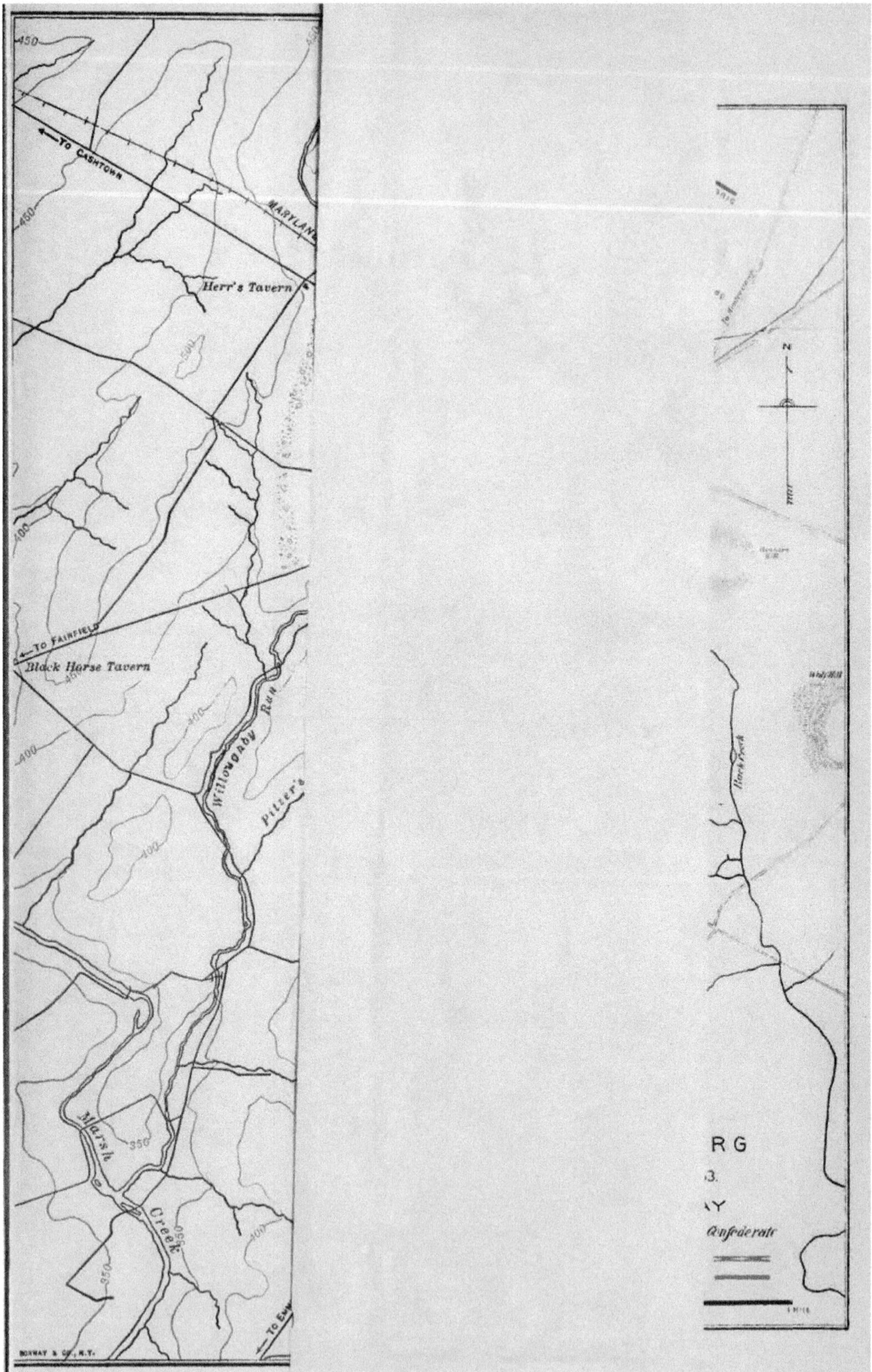

450
TO CASHTOWN
MARYLAND
Herr's Tavern
500
400
TO FAIRFIELD
Black Horse Tavern
Willoughby Run
Pitzer's
350
Marsh
Creek
N
R G
Confederate

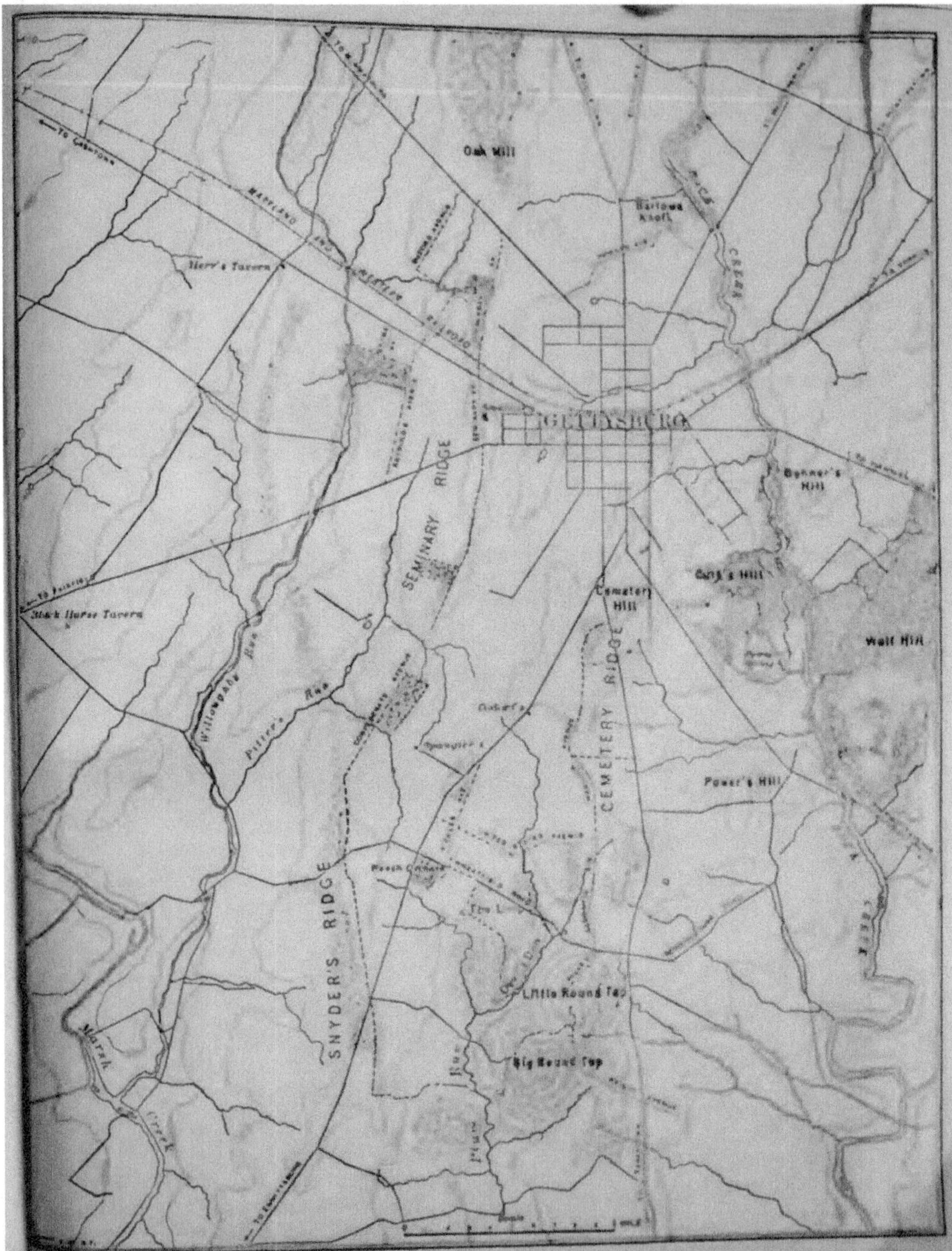

Oak Hill
GETTYSBURG
SEMINARY RIDGE
CEMETERY RIDGE
SNYDER'S RIDGE
Cemetery Hill
Culp's Hill
Wolf Hill
Power's Hill
Little Round Top
Big Round Top
Peach Orchard
Herr's Tavern
Willoughby Run
Pitzer's Run
Marsh Creek

From "MAINE AT GETTYSBURG."

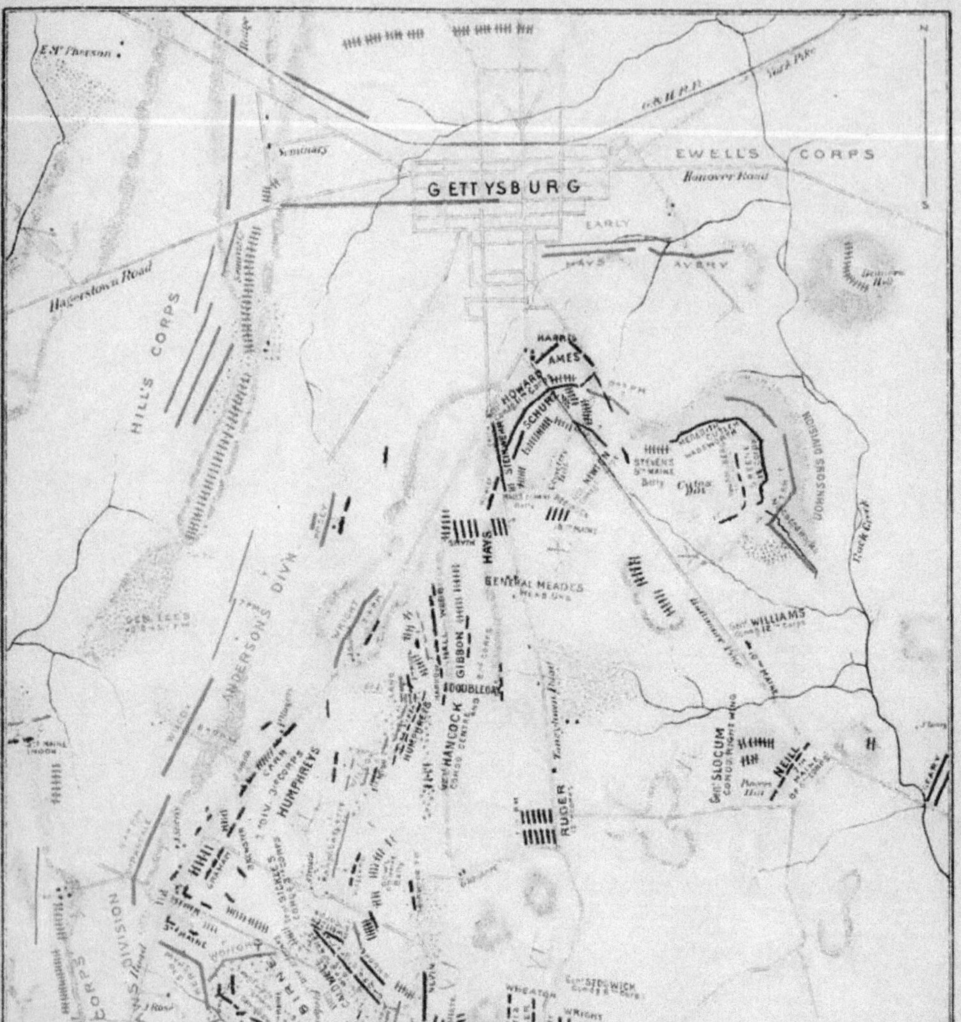

From "MAINE AT GETTYSBURG."

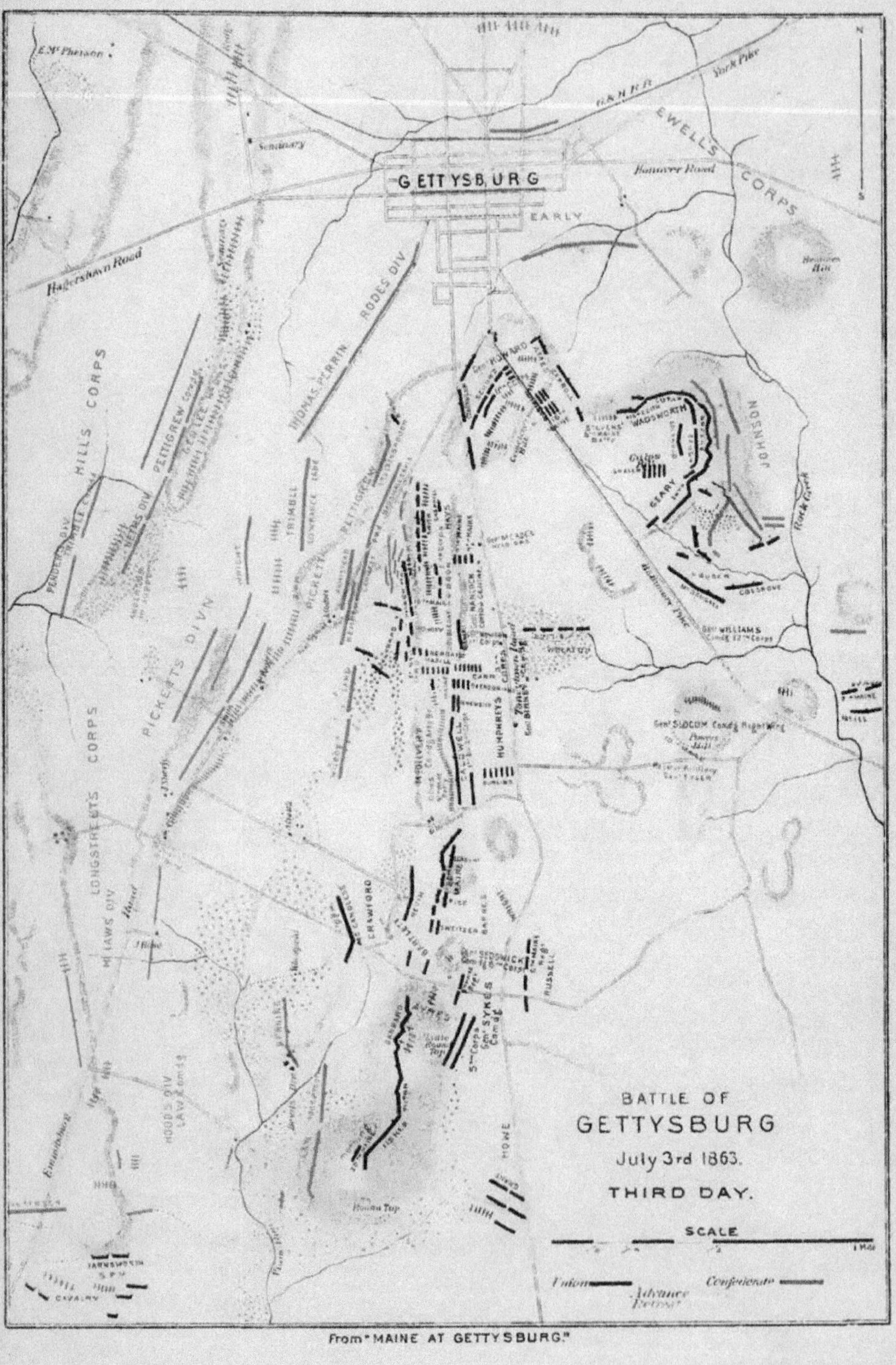

From "MAINE AT GETTYSBURG."

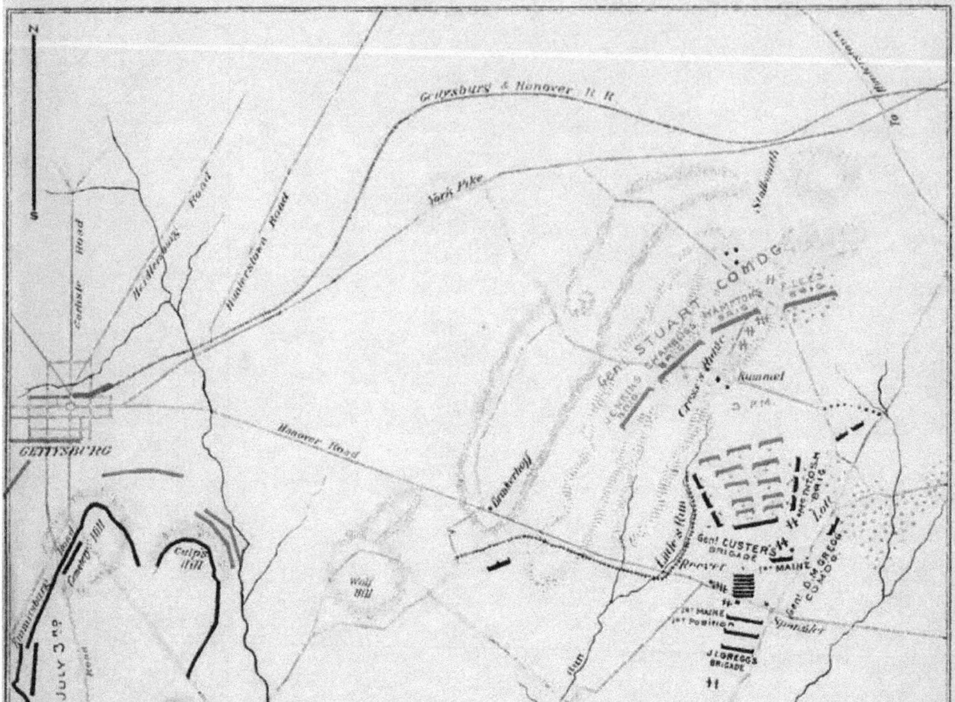

From "MAINE AT GETTYSBURG."